Advanced Placement Calculus with the TI-89*

Ray Barton
Olympus High School
Salt Lake City, UT

John Diehl
Hinsdale Central High School
Hinsdale, IL

Texas Instruments Incorporated
7800 Banner Drive, M/S 3918
Dallas, TX 75251

Attention: Manager, Business Services

Printed in the United States of America.

ISBN: 1-886309-27-2

We invite your comments and suggestions about this book. Call us at **1-800-TI-CARES** or send e-mail to **ti-cares@ti.com**. Also, you can call or send e-mail to request information about other current and future publications from Texas Instruments.

Visit the TI World Wide Web home page. The web address is: **education.ti.com**

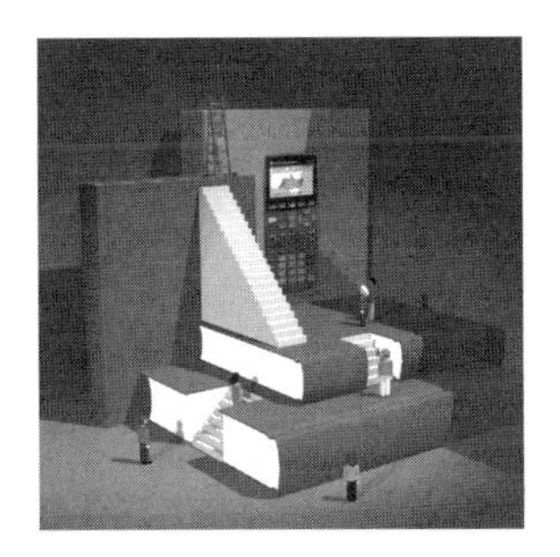

Contents

Preface

The TI-89 incorporates graphical and numerical features with a powerful computer algebra system that has the potential to dramatically alter how and what our students should learn. By incorporating this technology in the mathematics classroom, our students truly have a tool that facilitates a "three-fold" (numeric, graphic and analytic) approach to understanding and using mathematics.

Teachers can use the computer algebra capability of the TI-89 to create a lab setting where students discover concepts and theorems. When students discover a theorem they feel a sense of ownership and an interest in the proof of the theorem.

This book presents an introduction to the numeric, graphic and analytic features of the TI-89. It is our hope that as teachers and students become familiar with these features, they will experience the same excitement all mathematicians feel when a new idea is discovered.

We would like to thank the staff at Texas Instruments for their support, as well as Sally Fishbeck and Gary Luck for their valuable suggestions during the development of this book. We would also like to thank our families for their patience while we were writing.

— *Ray Barton*

— *John Diehl*

About the Authors

RAY BARTON teaches mathematics at Olympus High School in Salt Lake City. He received Utah's Presidential Award for Excellence in Mathematics Teaching in 1995. Ray presents workshops for the *Teachers Teaching With Technology* (T^3) program where he enjoys meeting teachers and investigating ways to enhance mathematics education.

JOHN DIEHL has been a high school Math teacher since 1977, working at Hinsdale Central High School in Illinois since 1980. He has been an instructor with the *Teachers Teaching with Technology* program since 1990 and has been an AP* Statistics consultant for the College Board since 1997. John received the Presidential Award in 1994. He has made numerous presentations about Texas Instruments graphing calculators on the national and international level. He served as the Texas Instruments visiting scholar in 1997-1998.

---

* AP and Advanced Placement Program are registered trademarks of the College Entrance Examination Board, which was not involved in the production of and does not endorse this project.

Chapter 1

Functions, Graphs, and Limits

In this chapter, you will use the TI-89 to create graphs and tables of values for functions. You then can use this graphical and numerical information to explore limits. You also will see how to evaluate limits directly.

Checking the mode settings

You will need certain mode settings to do the examples in this chapter. To check your settings:

1. Press [MODE] to see Page 1 of the settings. The settings you need are shown.

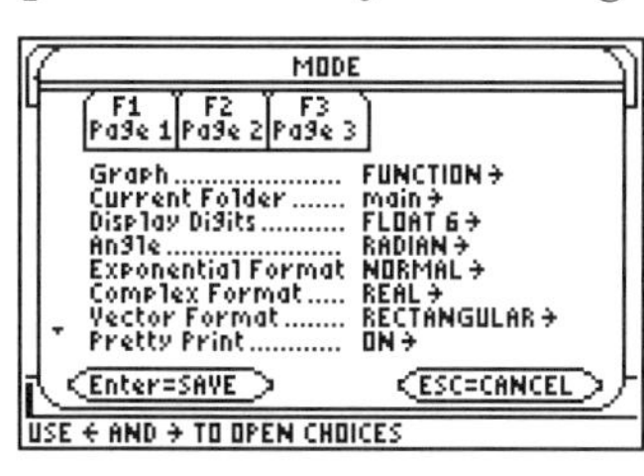

   To change a setting, first press ⊙ or ⊙ to highlight that setting.

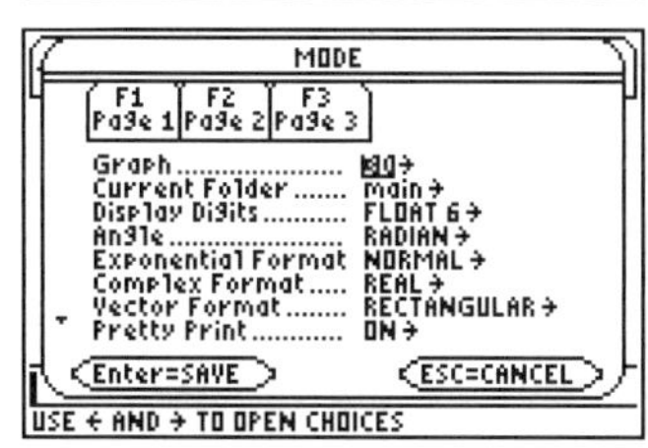

   Then press ⊙ to see the valid settings. Press the number of your choice.

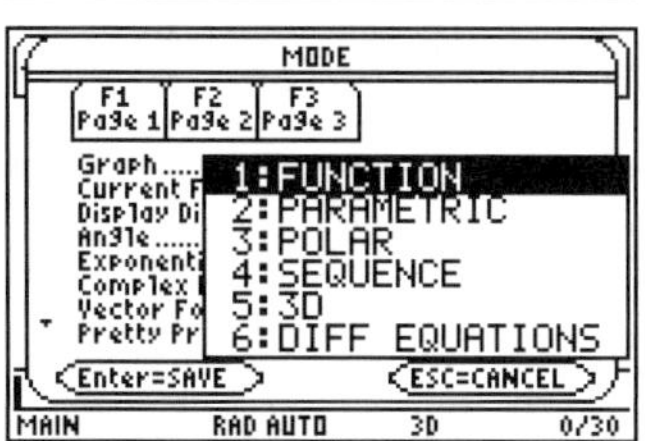

2. When the Page 1 settings are complete, press [F2] to check Page 2. Following the directions in step 1, make changes as needed. Note that certain items are dimmed. This simply indicates that they are not needed at this time. If you changed to a split screen, for example, more choices would appear.

   Use [F3] to check Page 3 in the same way.

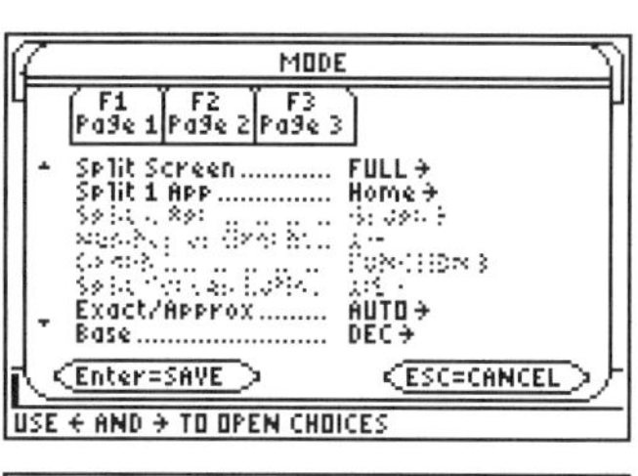

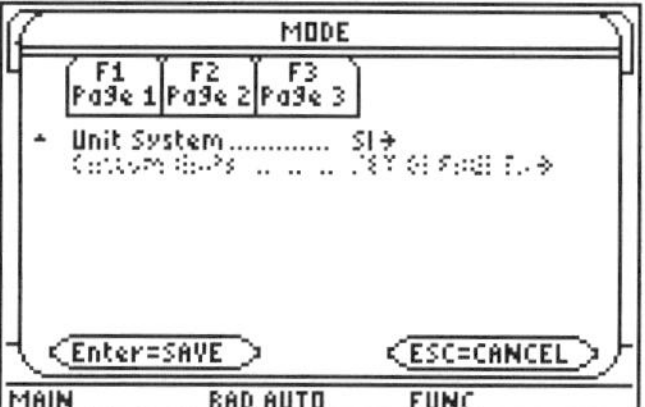

3. Press [ENTER] to save the new settings and exit from the MODE dialog box. If you did not return to the Home screen, press [HOME]. Press [F1] **Tools** and select **8:Clear Home** to clear the Home screen. If necessary, press [CLEAR] to clear the entry line.

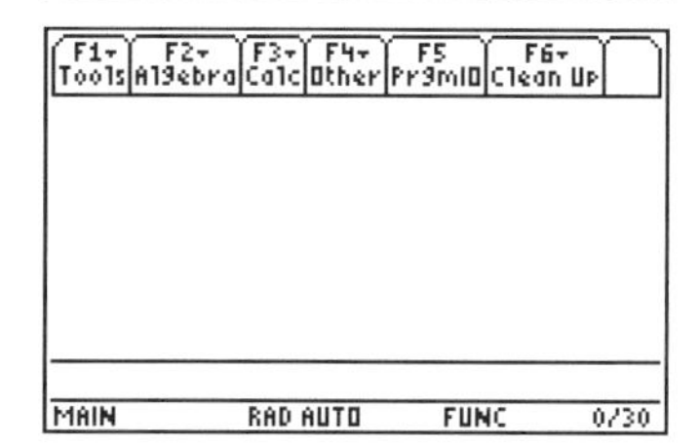

4. Press [2nd] [F6] **Clean Up** and select **2:NewProb** to clear variables and other settings.

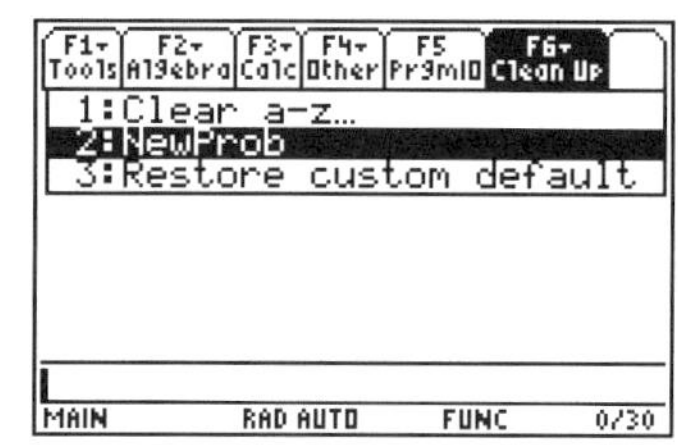

5. Press [ENTER].

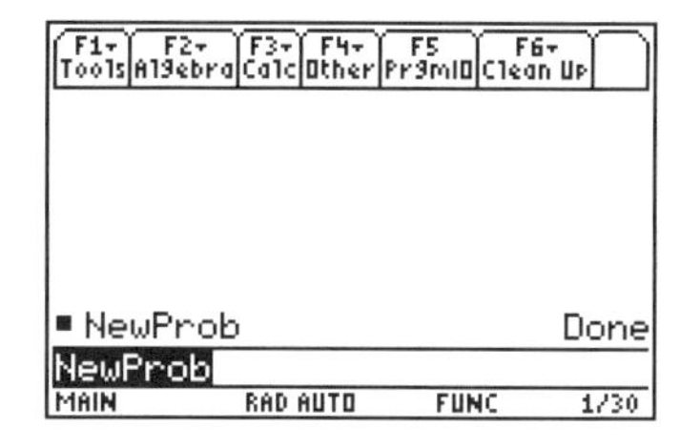

Example 1: Graphing a rational function

In this example, you will plot a rational function and explore function values on the graph screen and by table. The function is

$$y = \frac{2x^2 + 2x - 12}{x^2 - 4}.$$

What is y when $x = {}^{-}1$? What is the x intercept? What is x when $y=1$? Create a table of values for $x \in \{-4,-3,-2,-1,0,1,2,3\}$.

Solution

To find the answers, first enter the function in the Y= Editor. Use trace, zero, and intersection to compute function values. Then use the Table screen to enter the values for x.

Enter the function

1. To enter this function, press [◆] [Y=] to display the Y= Editor.

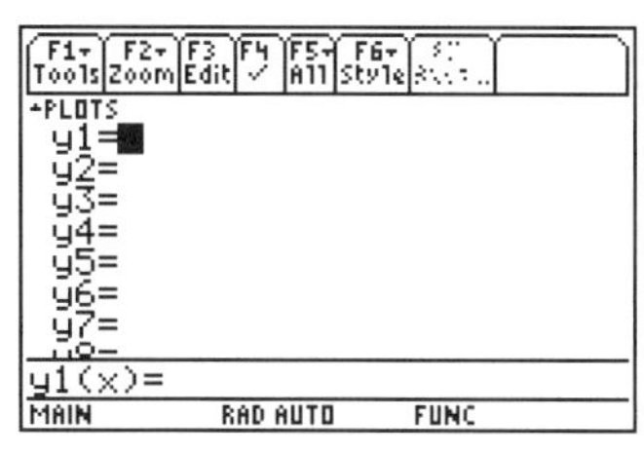

2. Define $y1(x)$. Be sure to use parentheses for the numerator and denominator. Use [^] to designate exponents. The function appears in the entry line.

   [(] **2X**[^] **2** [+] **2X**[−] **12** [)] [÷] [(] **X**[^] **2** [−] **4**[)]

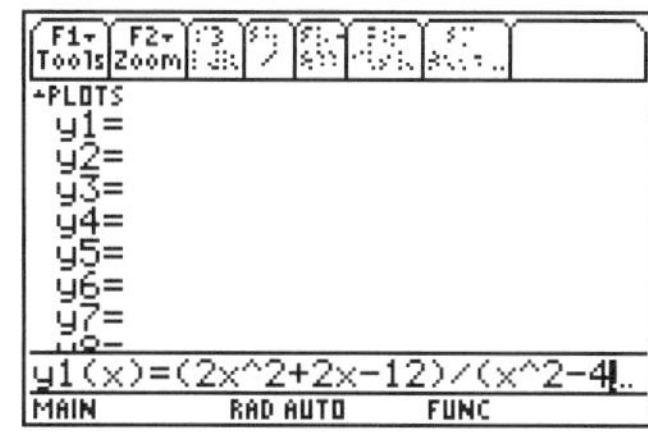

3. Press [ENTER].

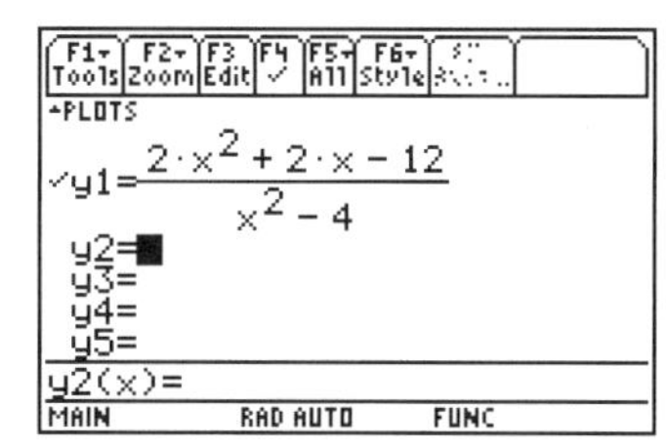

Define the viewing window

To define a viewing window, press [◆] [WINDOW] to display the Window Editor. Use ⊖ or [ENTER] to move through the choices and alter as needed to set the Window variable values as shown.

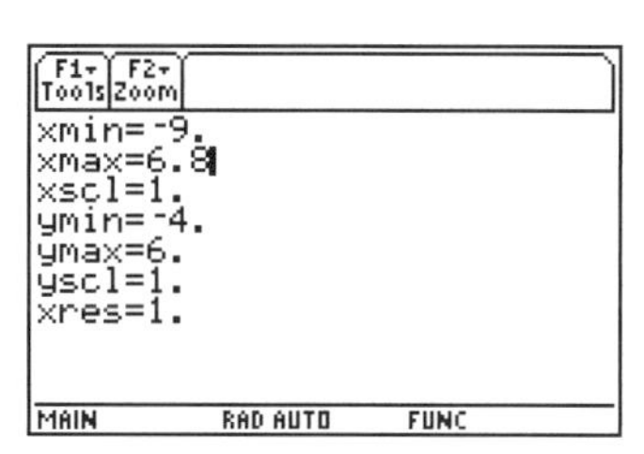

Graph

To plot the graph, press [◆] [GRAPH].

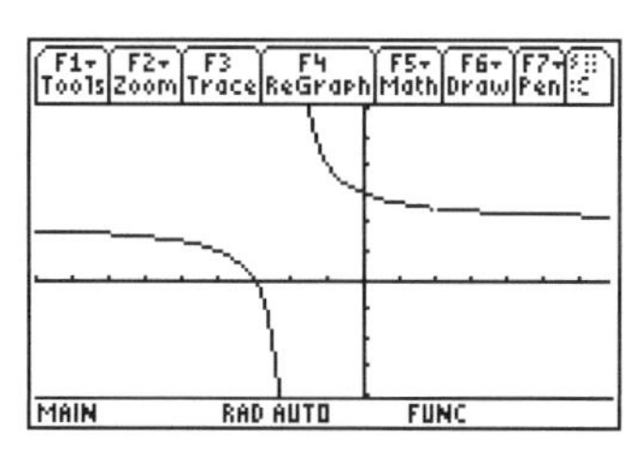

Compute function values

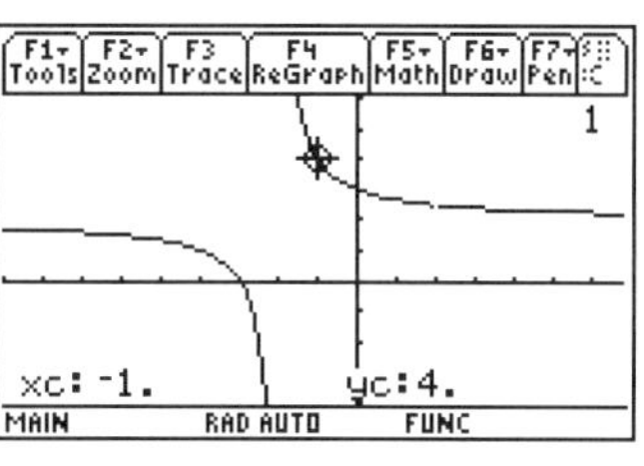

1. To trace the graph, press [F3] **Trace**. Use ◀ or ▶ to move the cursor along the graph. To move directly to a point, type the x-coordinate. For example, press [(-)] **1** [ENTER].

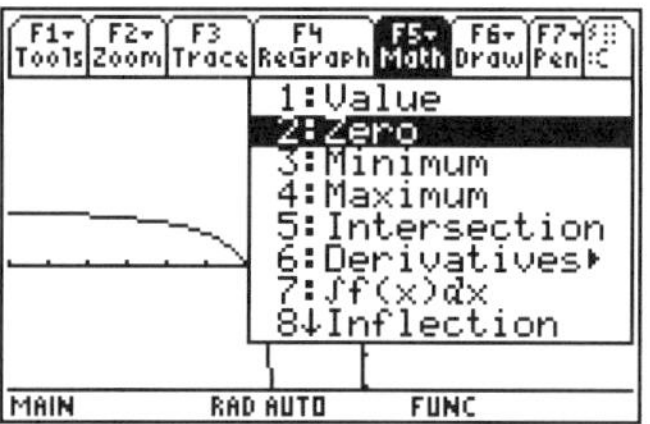

2. To compute an x-intercept, press [F5] **Math** and select **2:Zero**.

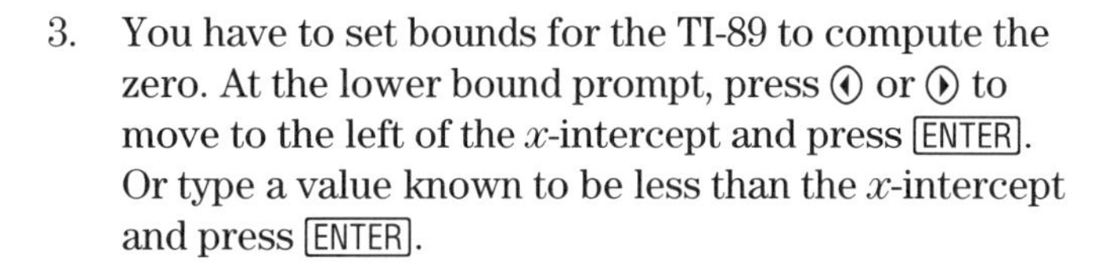

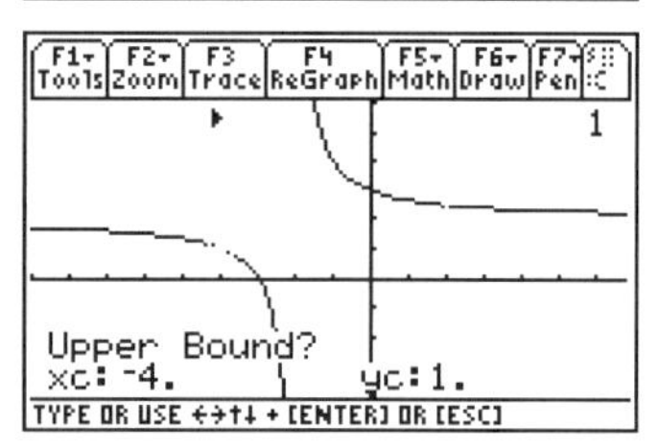

3. You have to set bounds for the TI-89 to compute the zero. At the lower bound prompt, press ◀ or ▶ to move to the left of the x-intercept and press [ENTER]. Or type a value known to be less than the x-intercept and press [ENTER].

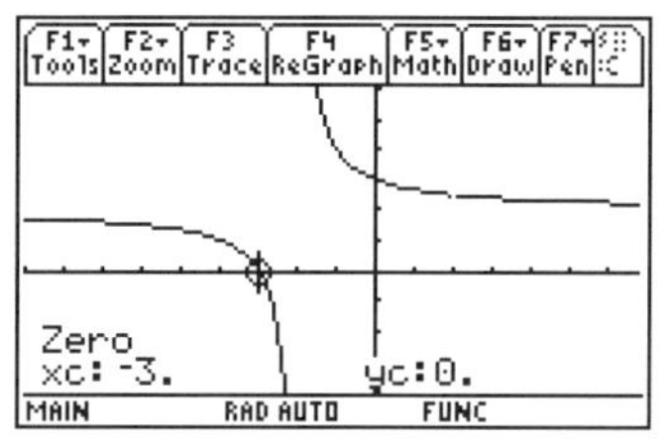

4. Similarly, enter an upper bound and press [ENTER]. The x-intercept of the graph zero of the function will be estimated.

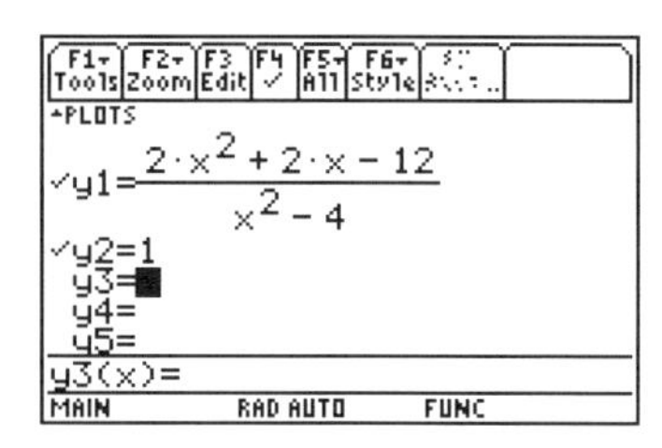

5. Often, you wish to compute an x-coordinate for a given value of y. One easy way to do this is to plot a horizontal line and compute the intersection point(s) of the original function and the horizontal line.

   To compute x when y=1, press [◆] [Y=] and enter a second function $y2$=1, the horizontal line. Press [ENTER].

6. Press [◆] [GRAPH] to graph the functions.

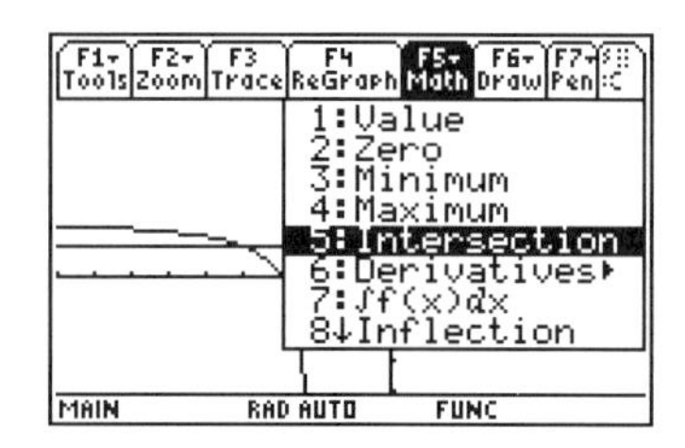

7. Press [F5] **Math** and select **5:Intersection**.

8. Press [ENTER] to select $y1$ as the first and curve. Press [ENTER] again to use $y2$ as the second curve. Press ◀ or ▶, or type values, to set the bounds. This works just like the zero math tool described previously.

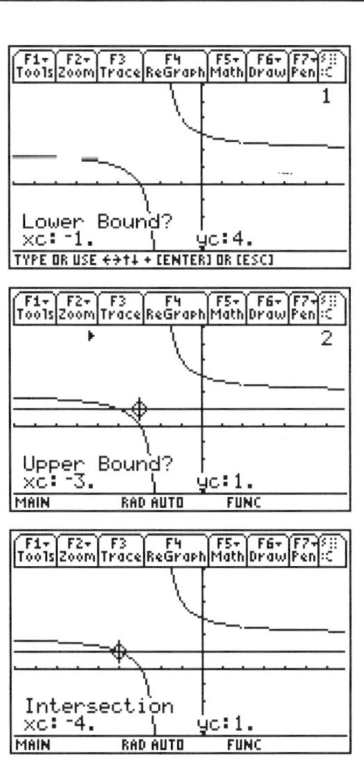

The coordinates of the intersection point are displayed.

Create a table of values

You will often wish to examine the behavior of a function at more than just a single point. The behavior can be examined both graphically and numerically. The numerical information is often best examined by a table of values.

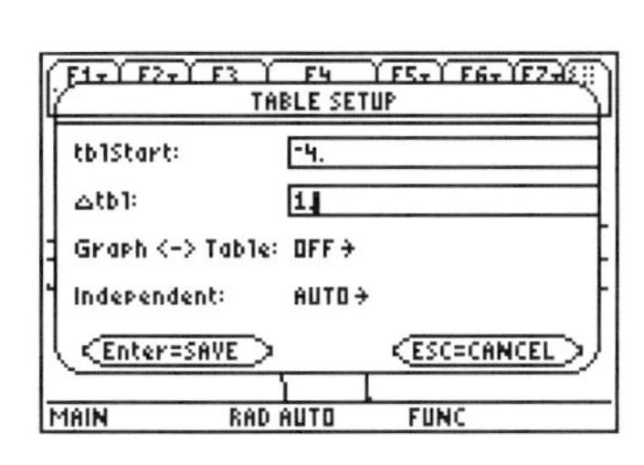

1. To create a table of values, first press ◆ [TblSet]. Enter a starting value (Δ**tbl start** =-4), and an increment (Δ**tbl**=1). Be sure to press [ENTER] to save the new values.

2. Your TI-89 will plot a graph and display values in the table for any selected function. Selected functions can be identified by the check mark at the left of the name in the Y= Editor. In this particular example, the values for $y2$ aren't of interest since they are all 1. You can deselect $y2$ so that the values do not appear in the table. To deselect the function $y2$, press ◆ [Y=] to return to the Y= Editor, place the cursor on $y2$ and press [F4] ✓. The check mark is removed, indicating a deselected function. It will not be graphed or display values in the table.

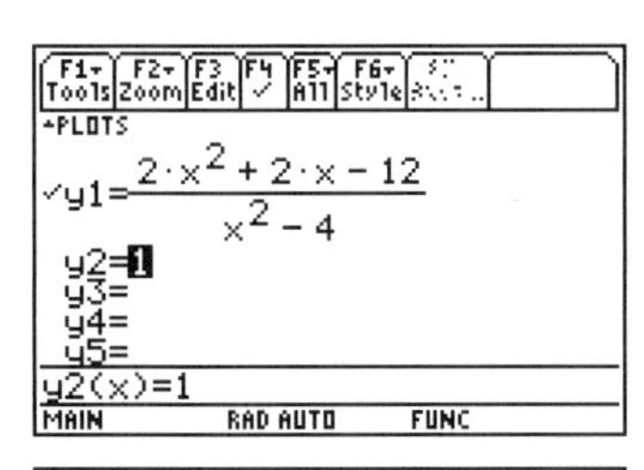

3. To see the table, press ◆ [TABLE].

x	y1
-4.	1.
-3.	0.
-2.	undef
-1.	4.
0.	3.

x= -4.
MAIN RAD AUTO FUNC

Example 2: Investigating limits

In this example, you will continue to investigate this function, particularly continuity, asymptotic behavior, and limits.

Given $f(x)=\dfrac{2x^2+2x-12}{x^2-4}$ (the same function as in Example 1) investigate the following limits:

$$\lim_{x \to 2} f(x)$$

$$\lim_{x \to -2+} f(x)$$

$$\lim_{x \to -2-} f(x)$$

$$\lim_{x \to \infty} f(x)$$

$$\lim_{x \to -\infty} f(x)$$

Solution

Investigate each limit graphically, numerically by a table of values, and with the **limit(** function.

Investigating graphically

1. Press ◆ [WINDOW] and set the same Window variable values as in Example 1 (at the right).

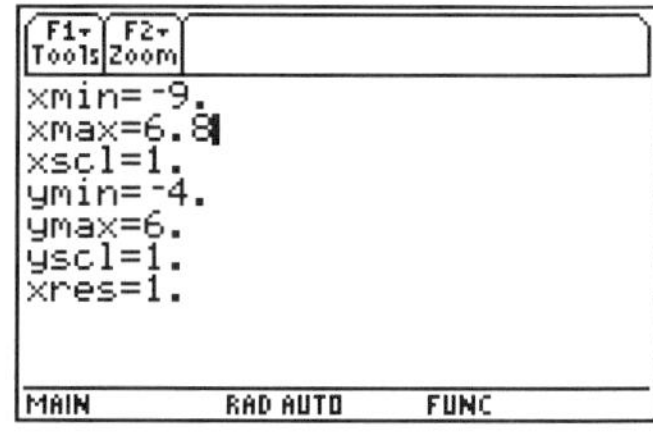

2. Press ◆ [GRAPH] to graph the function. You see evidence of a discontinuity at x=2. This is due to the "decimal" window where each pixel is .1 unit and **xres**=1 so that there is a pixel corresponding to x=2.

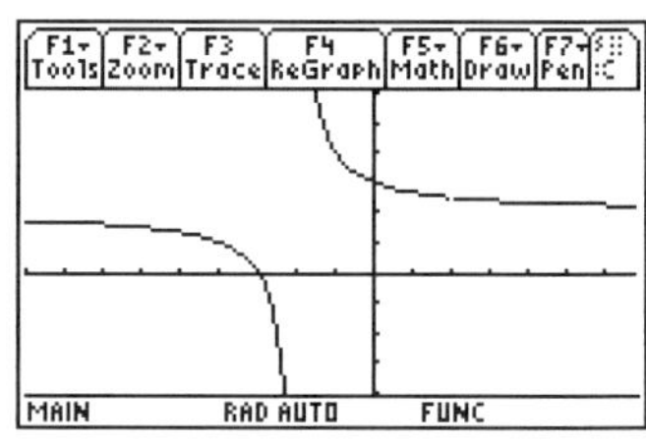

3. Return to the Window Editor, change **xmax** to 7, and graph the function.

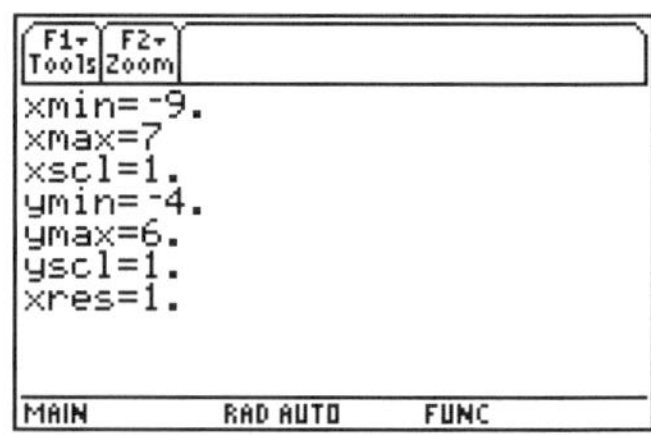

Notice that this minor change in the window settings affects the display. In this new window, there is no evidence of the discontinuity. It also makes a "false connection," resembling a vertical asymptote.

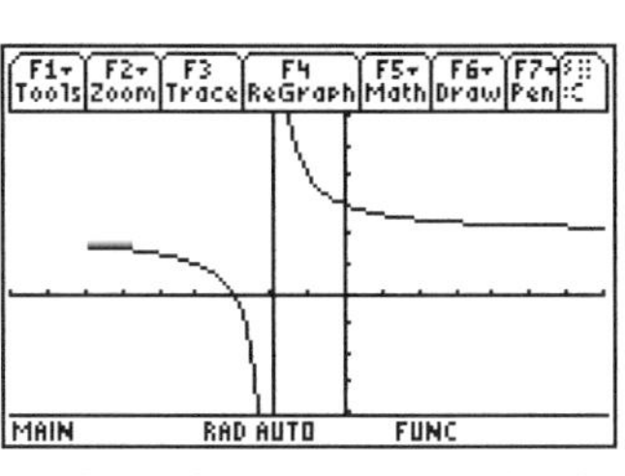

4. To investigate the limit as $x \to 2$, change **xmax** back to 6.8 and graph. Press [F3] **Trace** and trace near x=2.

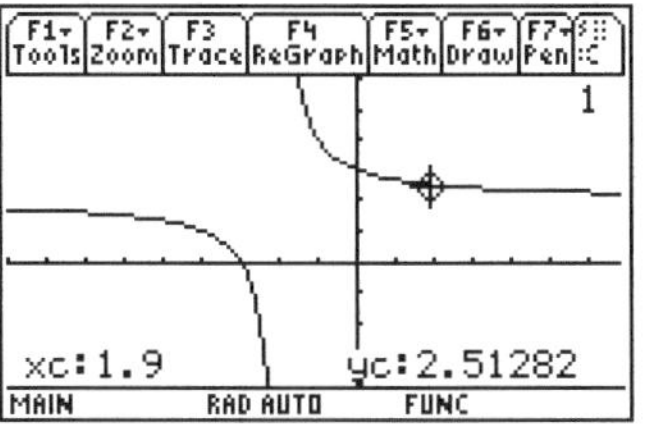

At x=2, no y-coordinate is displayed. Notice also that the display suggests a vertical asymptote near x= ⁻2 and a horizontal asymptote near y=2.

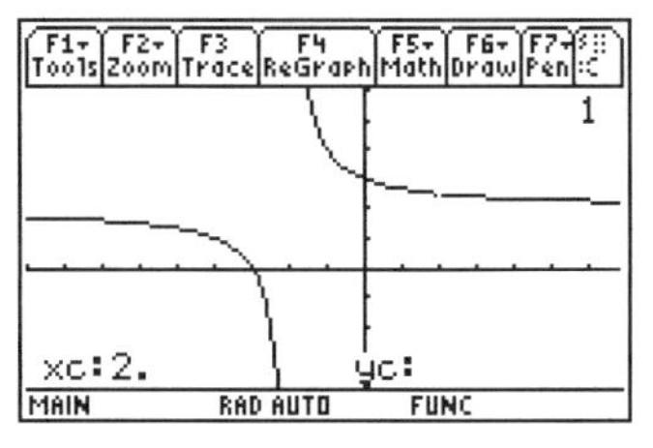

Investigating numerically

1. To investigate a limit numerically, press [◆] [TblSet].

   a. Use ⊖ to move the cursor to **Independent**.

   b. Press ⊙.

   c. Select **2:Ask** to change the selection.

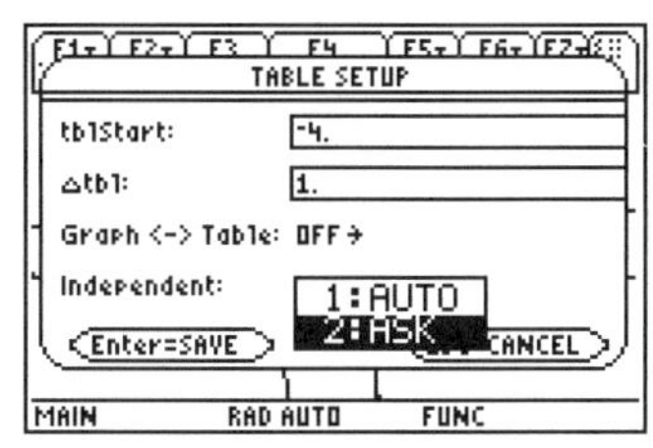

Be sure to press [ENTER] to save the change.

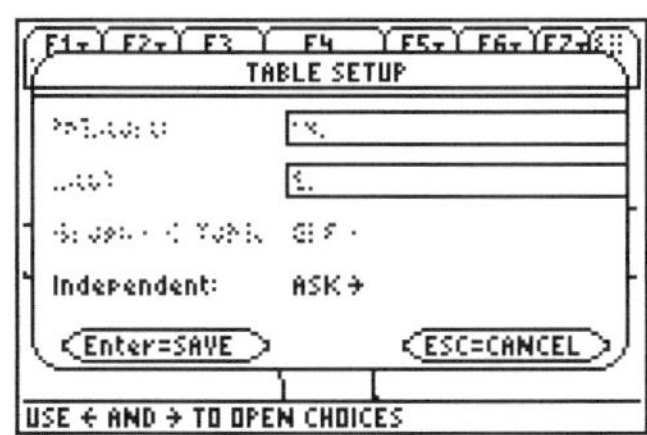

2. Now press [◆] [TABLE]. If necessary, press [F1] **Tools** and select **8:Clear Table**.

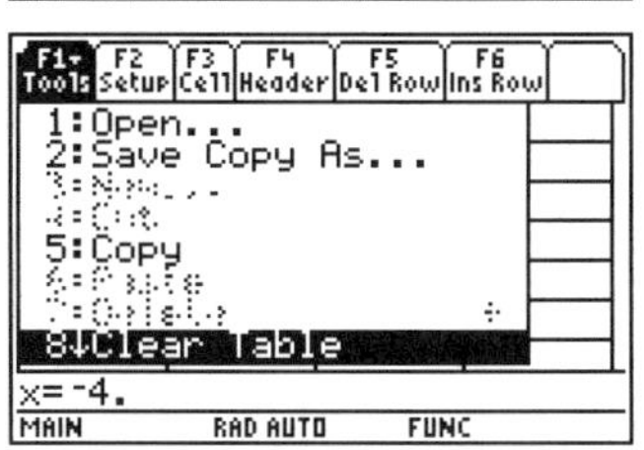

Press [ENTER] at the prompt to clear the table.

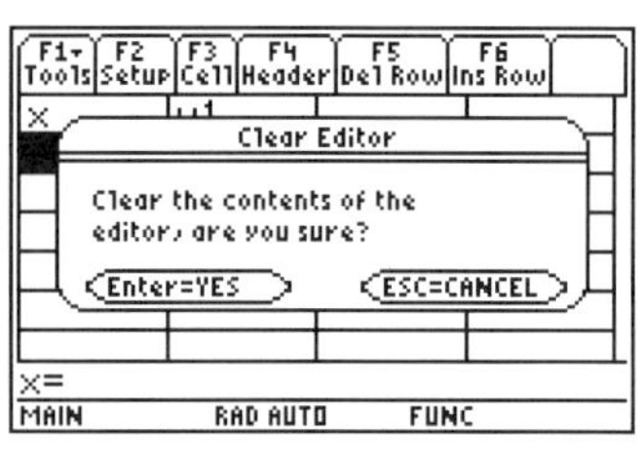

3. You can now enter values near 2 to investigate the limit. Press ⊙ to move to the next cell if you want the previous results to remain.

x	y1
1.9	2.5128
1.99	2.5013
1.999	2.5001
1.9999	2.5

y1(x)=2.5000125003125

4. Try values larger than 2 to investigate the right-hand limit.

x	y1
2.1	2.4878
2.01	2.4988
2.001	2.4999
2.0001	2.5

y1(x)=2.4999875003125

Using the limit command

The limit is the fundamental idea in a calculus course. With the TI-89, you can compute a limit as x approaches a particular value by entering the function, variable, and value.

1. Press [HOME] to return to the Home screen. Press [F3] **Calc** for the calculus menu.

2. Select **3:limit(** for the **limit** command. It will be pasted to the entry line on the Home screen.

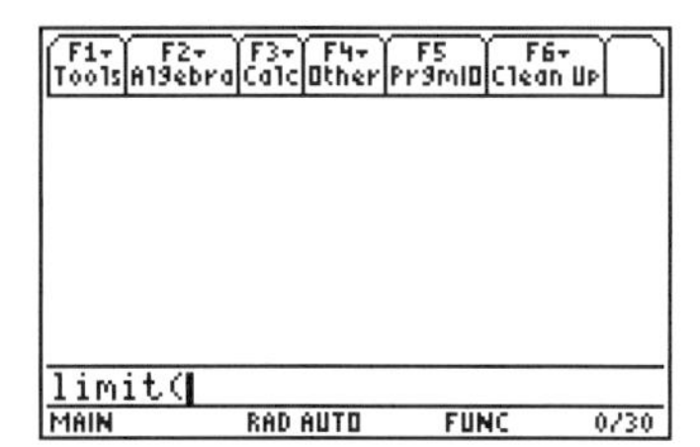

You want the function $y1(x)$ as the first argument of this command. You can type the function again. However, since you have already entered this function as $y1$, it is much easier to press [2nd] [RCL], type **y1** in the dialog box, and press [ENTER] twice.

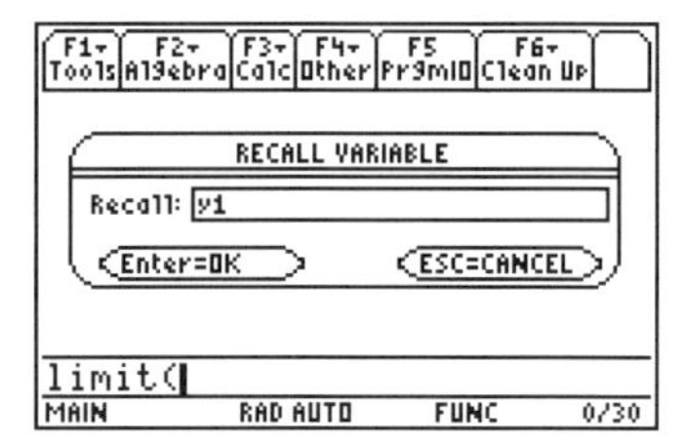

Then type the rest of the command.

[,] X [,] 2 [)]

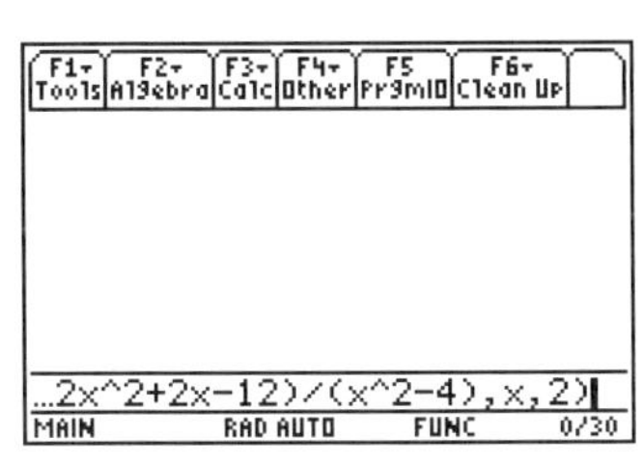

3. To compute the result, press [ENTER].

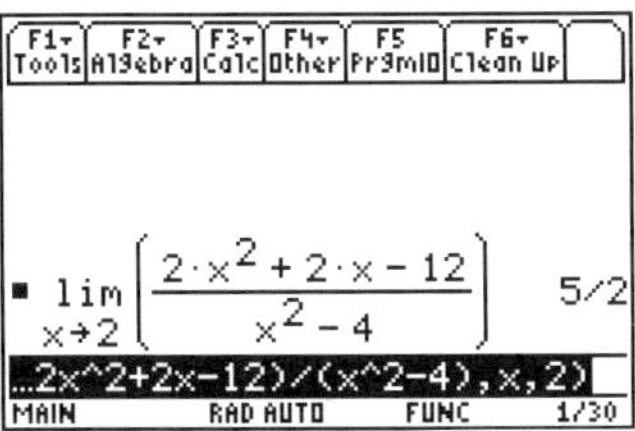

Alternately, you can use the function name, such as $y1$, in the **limit(** command.

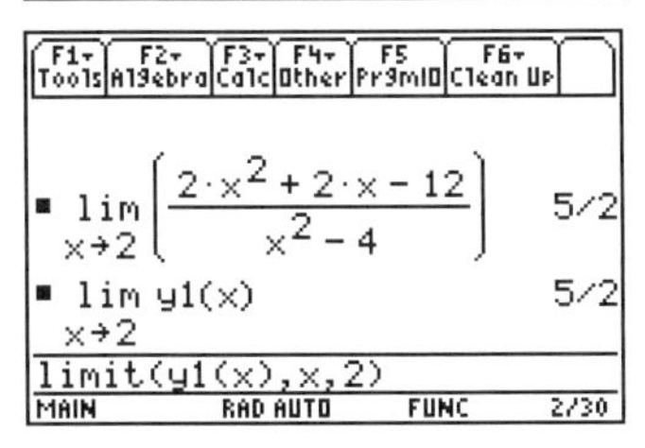

4. You also can find the **limit(** command in the CATALOG. Press [CATALOG]. The calculator assumes alpha mode, so just press **L**. Use ⊝ to point to the **limit(** command. Notice the help on the bottom line of the screen. It describes the parameters for the command. Press [ENTER] and proceed as described above.

Investigating other table values

1. To check that the table is still in the ASK setting, press ◆ [TblSet].

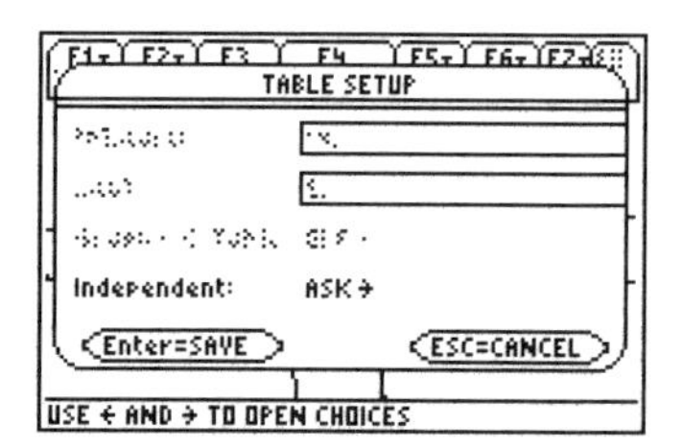

2. Press ◆ [TABLE] to return to the table .

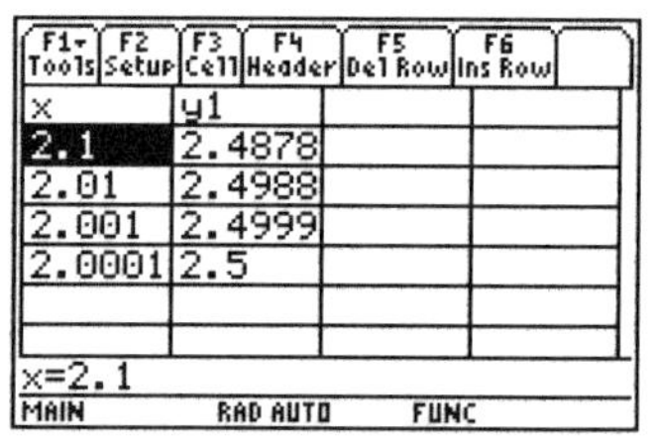

3. To clear the table, press [F1] **Tools** and select **8:Clear Table**. Then press [ENTER].

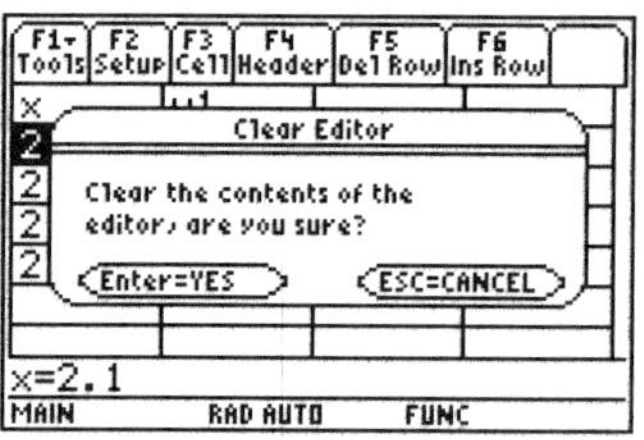

4. To change the format, press ◆ [I]. (You can also get to this screen by pressing [F1] **9:Format**.)

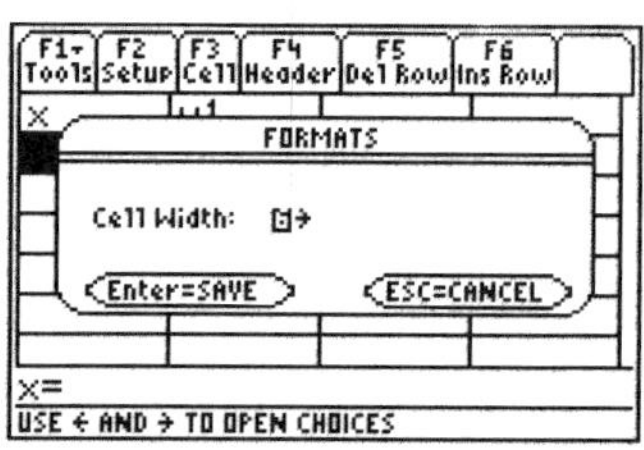

5. Change the cell width by using ▶, then ▼ to highlight **10**, and press [ENTER]. Press [ENTER] again to save the change.

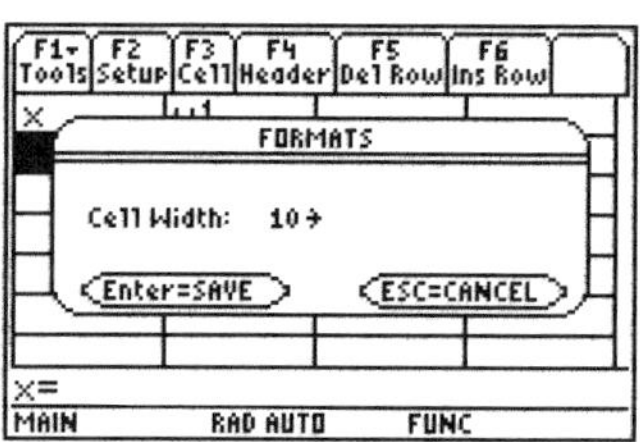

The columns should now be wider so that more digits can be displayed.

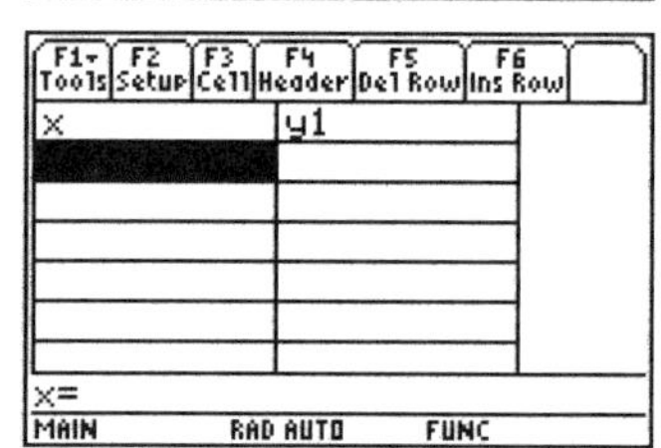

6. Type values near ⁻2 to explore the limit. Press [ENTER] to see the result, but remember to press ⊖ if you want to move to a new cell.

x	y1
⁻1.9	22.
⁻1.99	202.
⁻1.999	2002.
⁻1.9999	20002.
⁻1.99999	200002.

x= ⁻1.99999

7. Also check values on the other side of ⁻2.

x	y1
⁻2.1	⁻18.
⁻2.01	⁻198.
⁻2.001	⁻1998.
⁻2.0001	⁻19998.
⁻2.00001	⁻199998.

x= ⁻2.00001

8. Next, check values as $x \to \infty$.

x	y1
10.	2.16667
100.	2.01961
1000.	2.002
10000.	2.0002
100000.	2.00002

x=100000.

9. Finally, investigate as $x \to -\infty$.

x	y1
⁻10.	1.75
⁻100.	1.97959
⁻1000.	1.998
⁻10000.	1.9998
⁻100000.	1.99998

x= ⁻100000.

Entering more limit commands

It is also important to investigate limits when the results are infinite, or as x approaches positive or negative infinity.

1. To return to the Home screen, press [HOME]. Use ⊕ to highlight the previous **limit(** command. (If this command is not on your Home screen, just type the entire command as shown at the right.)

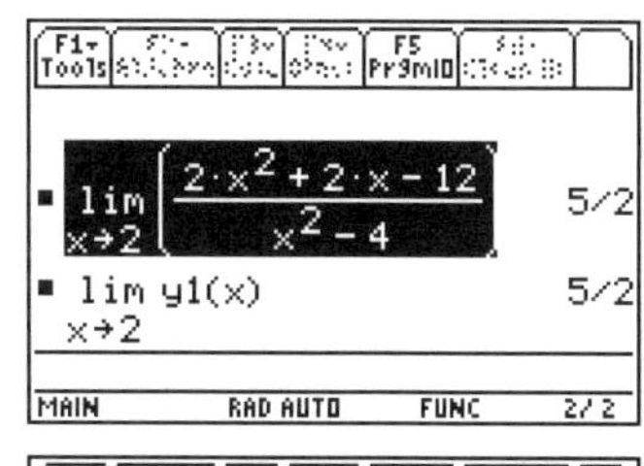

2. Press [ENTER] to copy the command to the entry line.

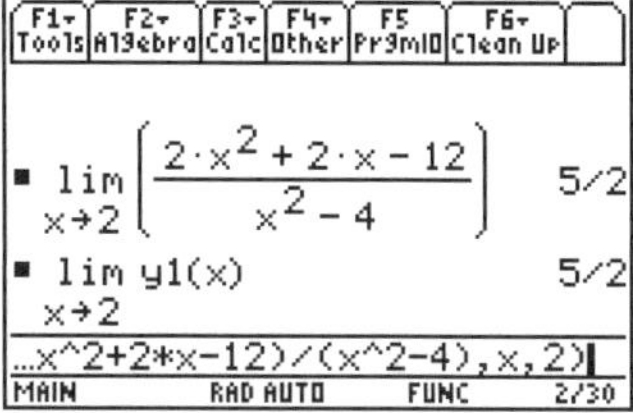

3. An insert cursor (vertical bar) will appear at the end of the line. Press ◀ twice to place the cursor in front of the 2.

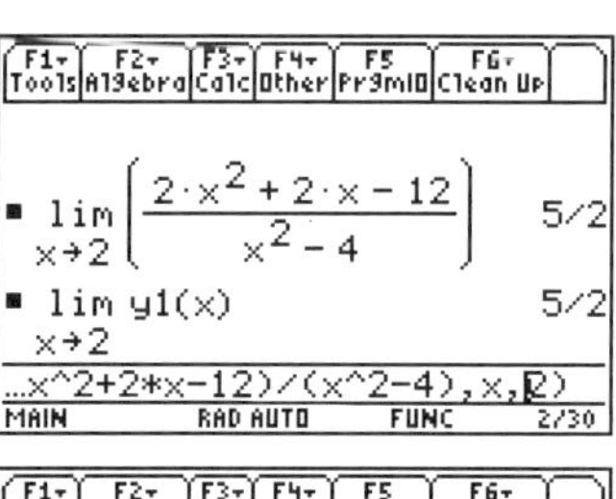

4. Press [(-)] to change the value to $^{-}2$.

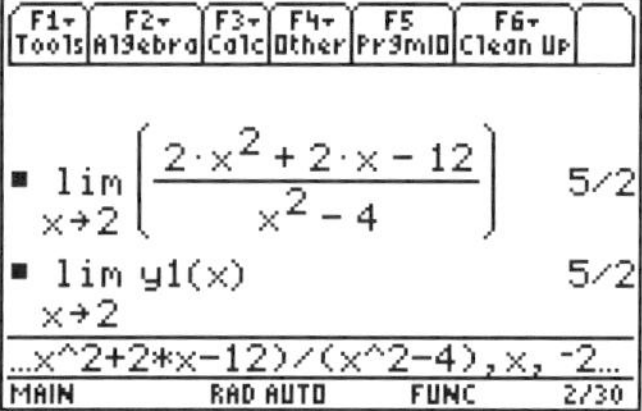

5. Press [ENTER] to evaluate.

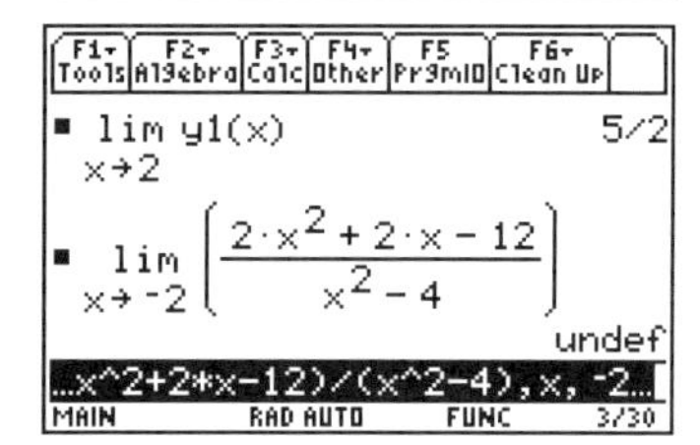

6. To investigate a one-sided limit, press ▶ to move to the end of the entry line. Then press ◀ to move the cursor between the 2 and the).

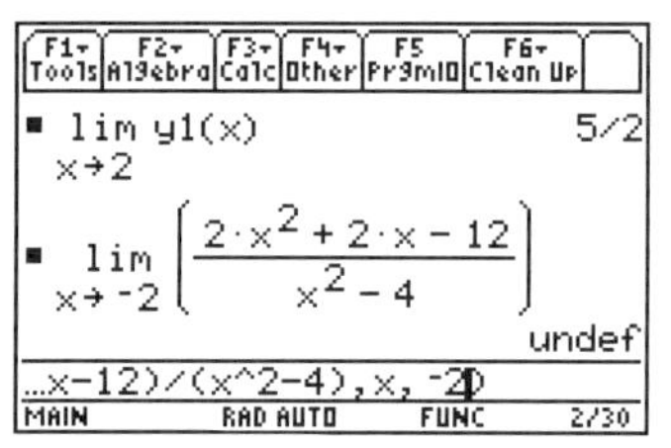

7. To evaluate a one-sided limit, use a fourth argument in the command. Type [,] **1**. (The **1** indicates a right-hand limit; any positive number will do.) Press [ENTER].

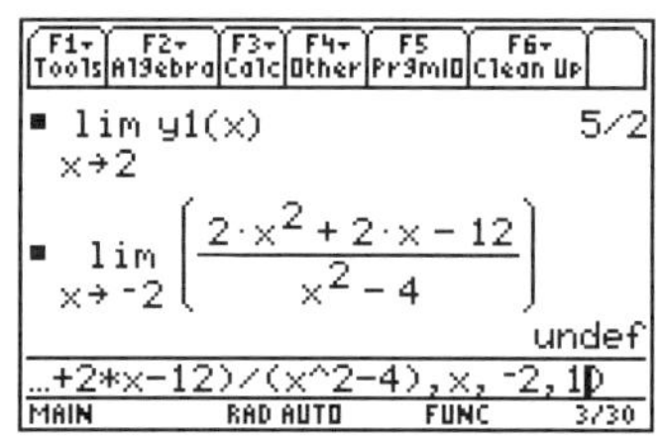

You can see that the result of infinity agrees with the asymptote you saw in the graph and in the numerical behavior of the table.

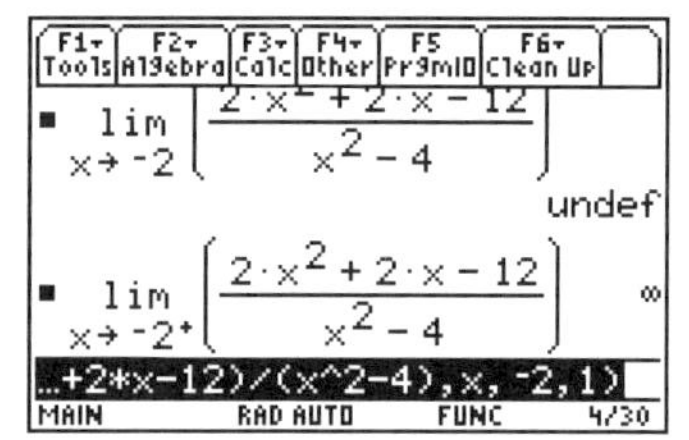

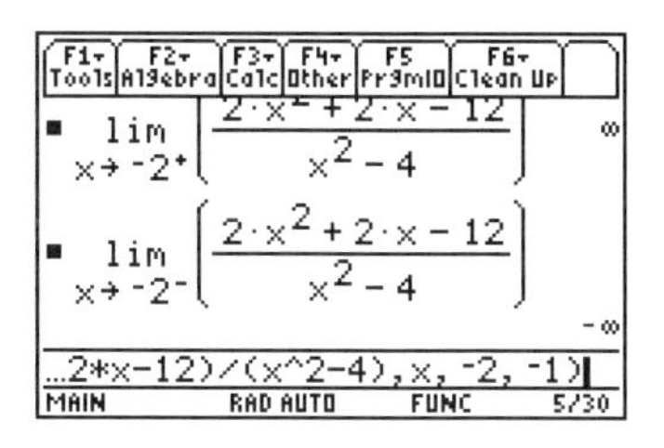

8. Investigate the left-hand limit in a similar manner. Note that entering ⁻1 created a left-hand limit, even though it is to the right of ⁻2. If the fourth argument is positive, the calculator computes a right-hand limit. If the fourth argument is negative, the calculator computes a left-hand limit, regardless of the value x approaches. The result of negative infinity again agrees with the graph, vertical asymptote, and the table.

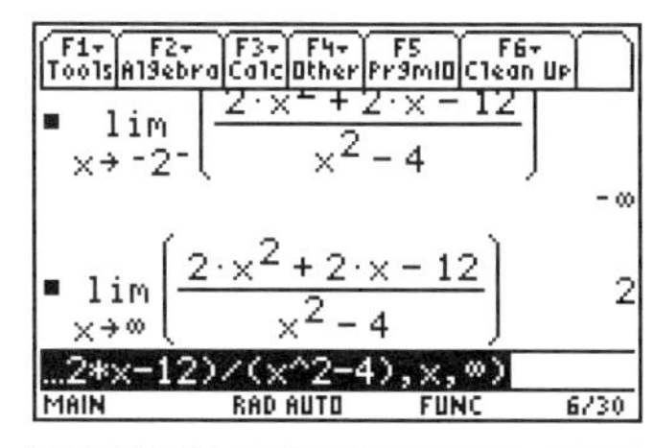

9. Press (▶) to move the edit cursor to the end of the entry line. Press [←] to remove the ⁻1 and the ⁻2 and press [◆] [∞] [)] to insert infinity. Press [ENTER] to see result. This result of 2 reinforces the horizontal asymptote and the numerical values in the table.

10. Finally, investigate the limit as $x \to -\infty$.

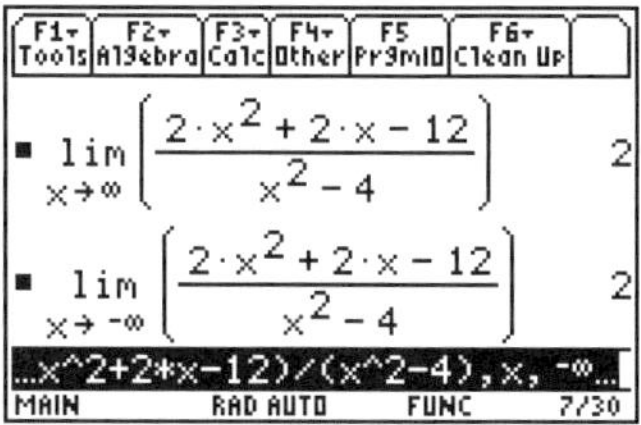

Exercises

Given $g(x) = \dfrac{2+x-x^2}{x^2-1}$

1. Graph $g(x)$.
2. What is $g(3)$?
3. What is the y-intercept?
4. What is the x-intercept?
5. What is x if $g(x)=3$?
6. Create a table of values for $x \in \{-3,-2,-1,0,1,2,3\}$.
7. What is $\lim_{x \to -1} g(x)$?
8. What is $\lim_{x \to 1} g(x)$? $\lim_{x \to 1+} g(x)$? $\lim_{x \to 1-} g(x)$?
9. What is $\lim_{x \to \infty} g(x)$? $\lim_{x \to -\infty} g(x)$?

Chapter 2

Differentiation

In this chapter, you will explore estimated and exact results for derivatives, tangent lines, implicit differentiation, and symbolic differentiation.

Example 1: Finding slope and the tangent line

In this example, you will use **avgRC**, the average rate of change command, and **nDeriv**, the symmetric difference quotient. These commands introduce difference quotients and how they are used to estimate slope. The limits of these difference quotients then define the derivative.

Given

$$f(x) = \sqrt{x^2 + 1},$$

estimate the slope of the curve when x=1. Determine the equation of the tangent line.

Solution

Use the traditional difference quotient (average rate of change) and the symmetric difference quotient (numerical derivative) to estimate the slope. Also, use a tangent command to draw the line and display the equation.

Calculating average rate of change

1. If needed, press [HOME] to return to the Home screen. You will now define a function. This definition will remain in the calculator's memory until it is cleared. Press [F4] **Other** and select **1:Define.**

2. Define the function.

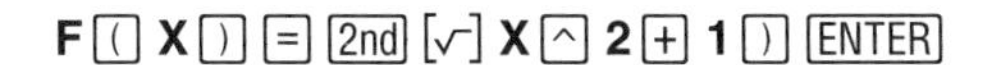

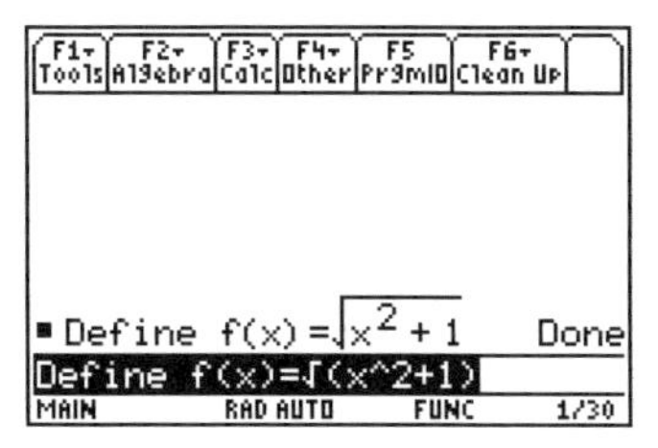

3. Press [CATALOG]. If needed, press **A** to move to commands that begin with A. Press [2nd] ⊖ to "page down" to **avgRC(**. Press [ENTER].

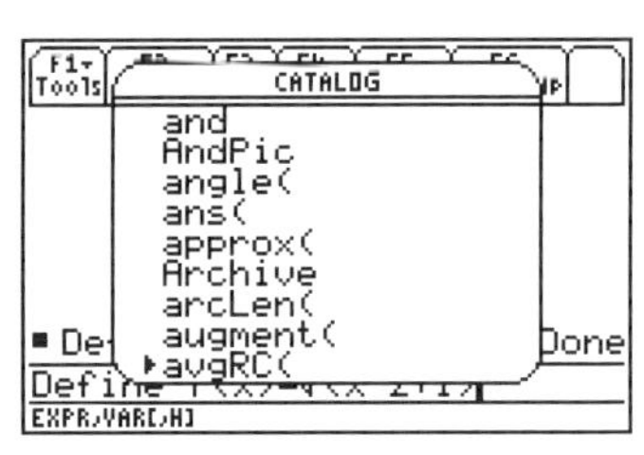

4. Complete the command.

   F [(] **X** [)] [,] **X** [)] [|] **X** [=] **1** [ENTER].

   Note that the default increment for the average rate of change is .001. Thus, the slope of the line through the points with x-coordinates 1 and 1.001 has been computed and is displayed. It is an estimate of the slope of the tangent line, or the slope of the curve, at the point x=1.

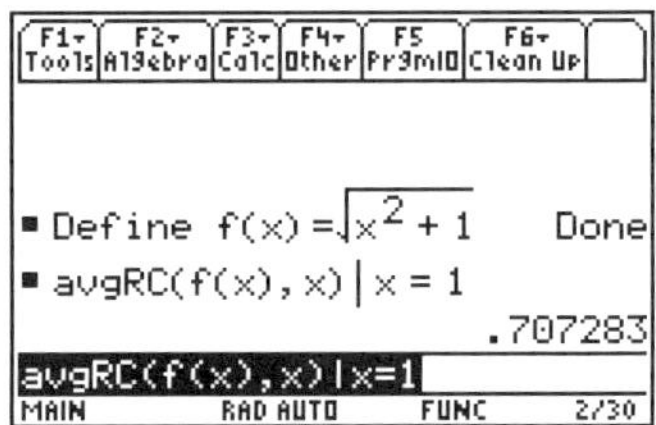

5. You can use a different increment if you enter it as the fourth argument of the **avgRC(** command. To use .0001, for example, press ⊙. The cursor is now at the right of the expression. Press ⊙ five times to move between the second x and the right parenthesis. Press [,], type **.0001** and press [ENTER].

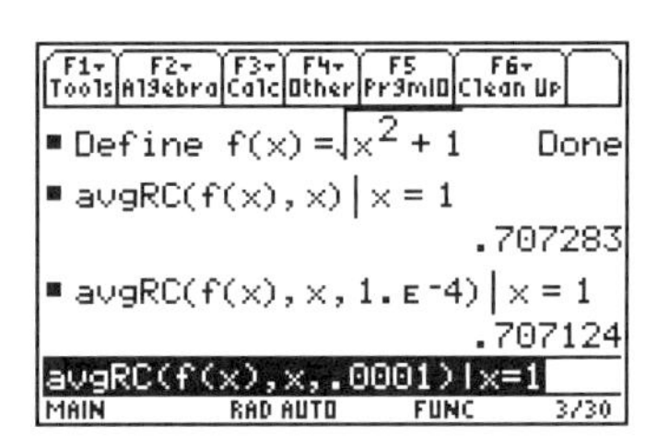

   This time the slope of the line through the points with x-coordinates 1 and 1.0001 has been computed and displayed. It may be a better estimate of the slope of the tangent line. Also, note that the increment value can be entered as a negative number, so that the slope line may be computed using a point to the left of the given point.

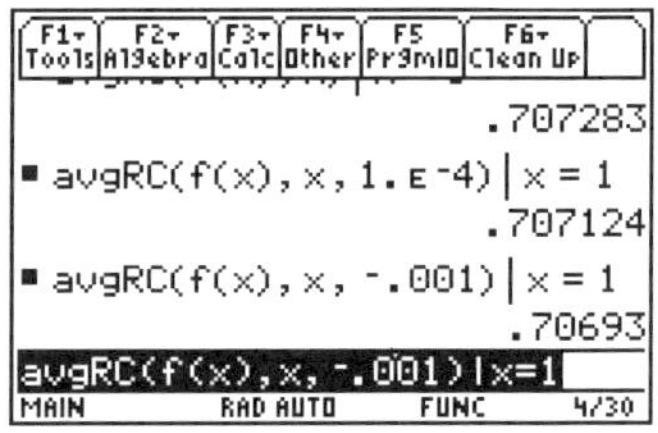

6. To gain some insight into the **avgRC(** command, you can use an undefined function such as $g(x)$. You want to be sure that it is not currently defined; however, you may not want to clear other variables such as the function f that was just defined.

   The best way to clear a few specified variables is by a **DelVar** command. To guarantee that g and h are undefined, press [F4] **Other**, select **4:DelVar**, enter **G** [,]**H** and press [ENTER].

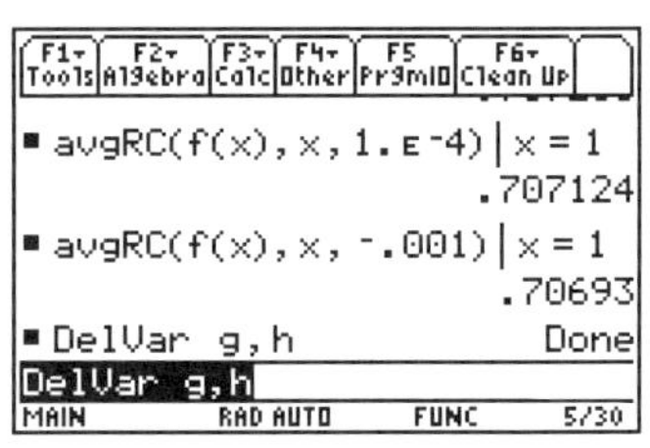

   Now press [CATALOG] to select **avgRC**. The selection arrow will still be at **avgRC** if you haven't done anything else in the CATALOG since step 3. Complete the command.

   G [(] **X** [)] [,] **X** [,] **H** [)] [ENTER]

   You now have an average rate of change for an arbitrary function with an arbitrary increment.

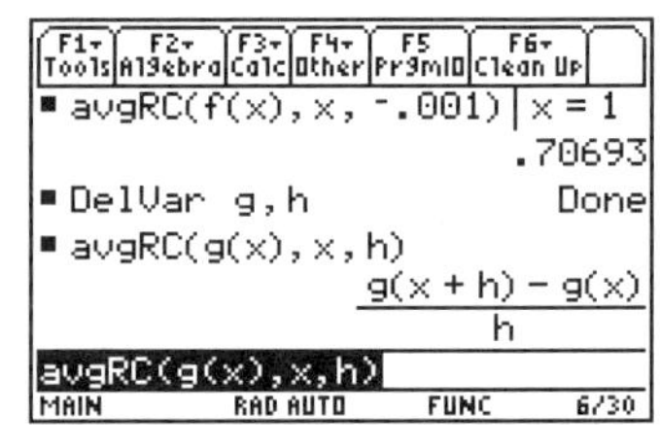

Finding the numerical derivative

If you compute an average rate of change for one positive and one negative value for h, you will often get one value that is an overestimate and one value that is an underestimate. You can average these results for another estimate. The same result can be obtained directly using the symmetric difference quotient, often referred to as the *numerical derivative*. It has been used more often recently since graphing calculators began using it for numerical differentiation estimates.

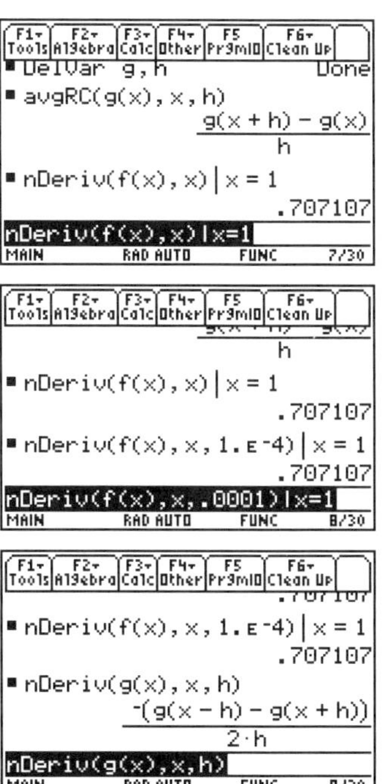

1. To try the same problem with a symmetric difference quotient, press [F3] **Calc** and select **A:nDeriv(**. Complete the command.

   F [(] **X** [)] [,] **X** [)] [|] **X** [=] **1** [ENTER].

   Again, the default increment is .001.

2. To try a smaller increment, insert **.0001** as the third argument, just as you did for **avgRC(**, and press [ENTER].

3. To gain some insight into the **nDeriv(** command, use the command on $g(x)$ with increment h as you did for **avgRC(**.

Drawing the tangent line

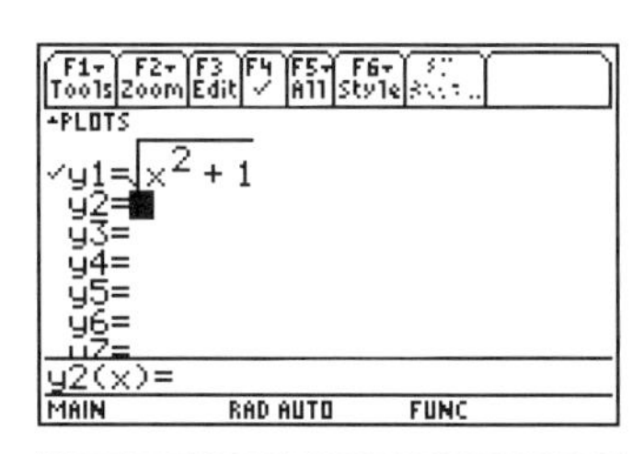

1. To enter the function for graphing, press [◆] [Y=]. If necessary, press ⊖ or ⊖ to move the cursor to $y1$. If there is an old equation, press [CLEAR]. As you type the function, it appears on an entry line near the bottom of the screen.

   [2nd] [√] **X** [^] **2** [+] **1** [)] [ENTER].

2. To set a convenient viewing window and center the origin, press [F2] **Zoom** and select **4:ZoomDec**.

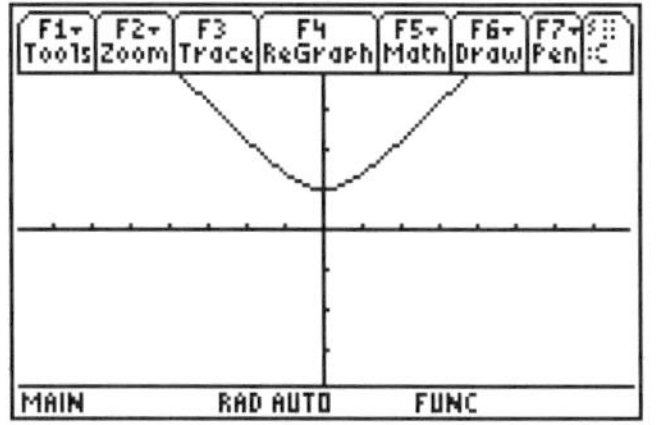

3. Compute the slope at a point on the graph. First, press [F5] **Math** and select **6:Derivatives**, and then select **1:dy/dx.** Now you are prompted for the x-coordinate. You can use the arrow keys, but it is easier just to type a value. For example, to input 1 as the x-coordinate, press **1** [ENTER]. The estimated value of the derivative is displayed.

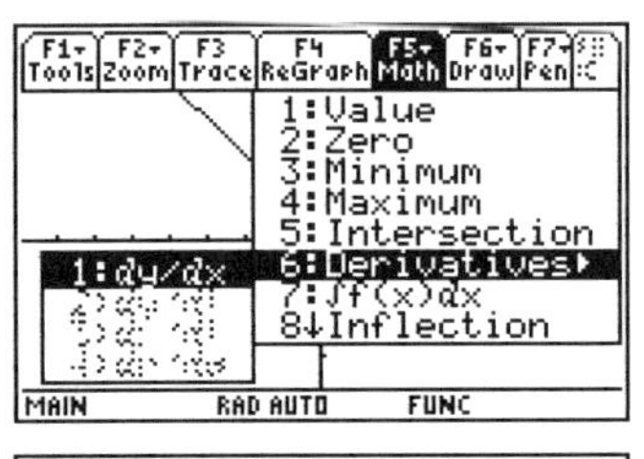

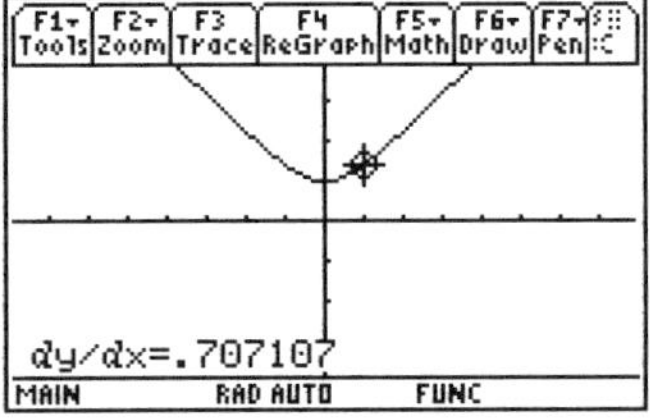

4. To plot a tangent line and display the equation, press [F5] **Math** and select **A:Tangent.** Enter **1** for the x-coordinate. The tangent line and its equation are displayed.

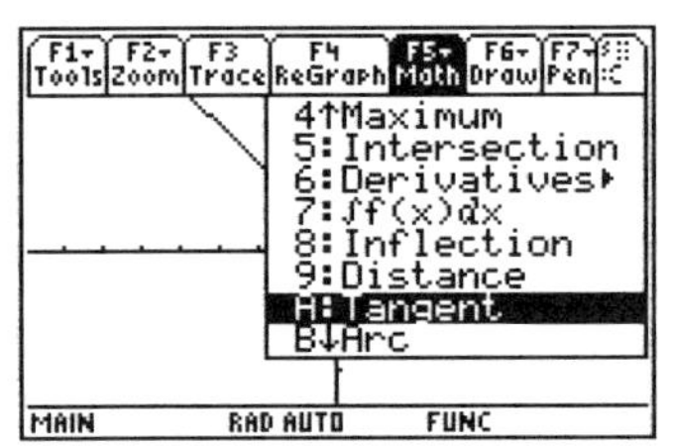

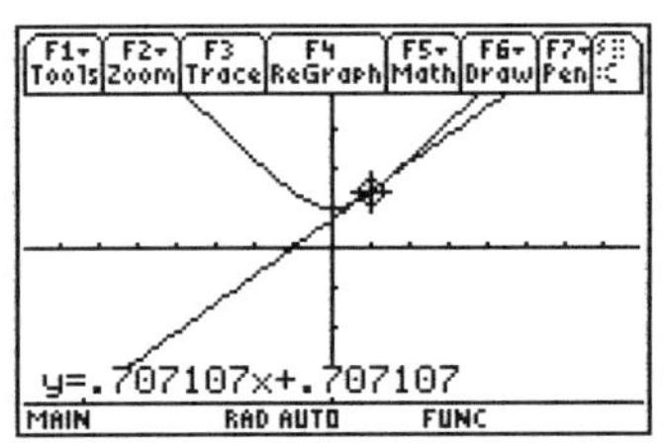

Example 2: Exact derivatives

Each of the methods so far have used an estimation method for the slope. With the TI-89, you can compute exact derivatives as well as the exact value for the slope at a given point.

Given the same function,

$$f(x) = \sqrt{x^2 + 1},$$

find $f'(x)$, $f''(x)$, and the exact value of $f'(1)$.

Solution

Explore the derivatives with limits and the built-in derivative command.

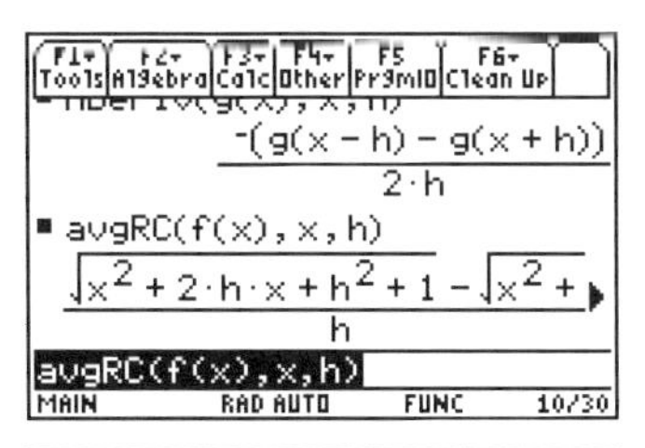

1. To explore with the traditional difference quotient, press [CATALOG], select **avgRC(**, and press [ENTER]. Complete the command to display the traditional difference quotient with increment h for this function.

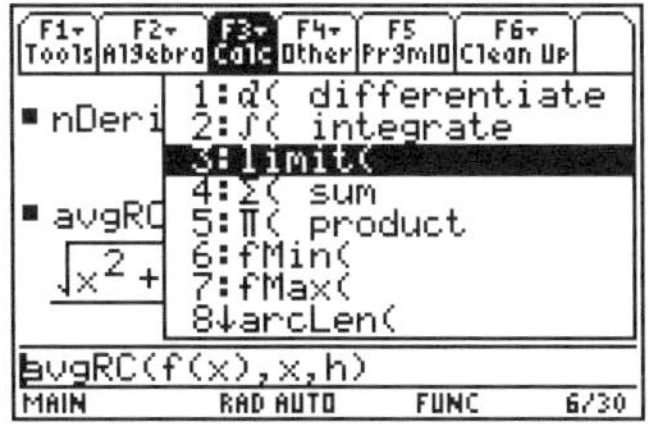

2. Of course, the derivative is the limit as h approaches zero of the expression you have just computed. To evaluate this limit, press ◀ to move the cursor to the beginning of the command. Then press [F3] **Calc** and select **3:limit(**. This will paste the **limit(** command at the beginning of the command.

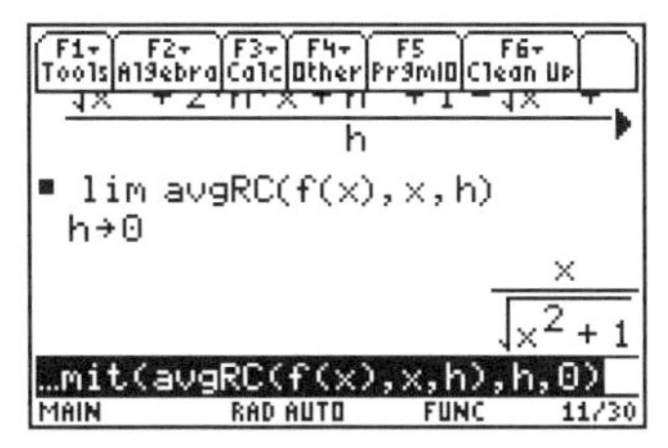

3. Now finish the command with the limit variable and value by pressing [2nd] ▶ to move the cursor to the end of the command line. Press [,] **H** [,] **0** [)] [ENTER]. The actual derivative is displayed, and it has been computed by the definition of the derivative.

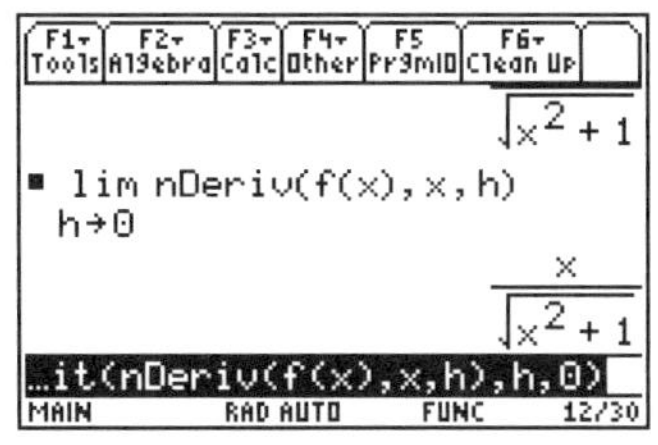

4. You also can compute a derivative as a limit of the symmetric difference quotient. Enter the command:

   [F3] **3:limit(** [F3] **A:nDeriv(** **F** [(] **X** [)] [,] **X** [,] **H** [)] [,] **H** [,] **0** [)] [ENTER]

   The actual derivative is displayed, and it has been computed as a limit of the symmetric difference quotient.

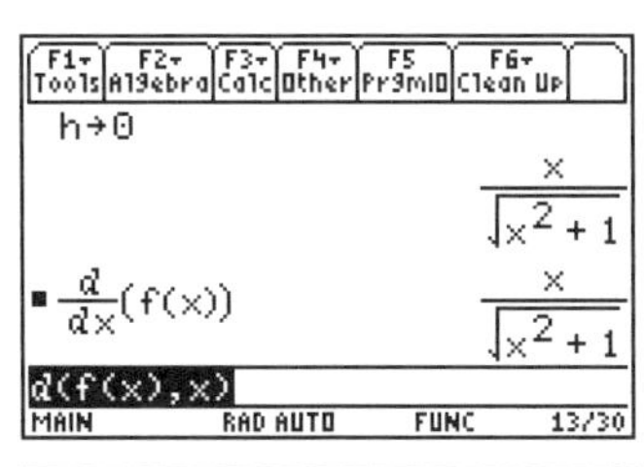

5. Although the limits computed above are important for the concept of the derivative, they are not the easiest way to compute the derivative. Use the built-in derivative command for the function *f(x)*.

   [2nd] [*d*] **F** [(] **X** [)] [,] **X** [)] [ENTER]

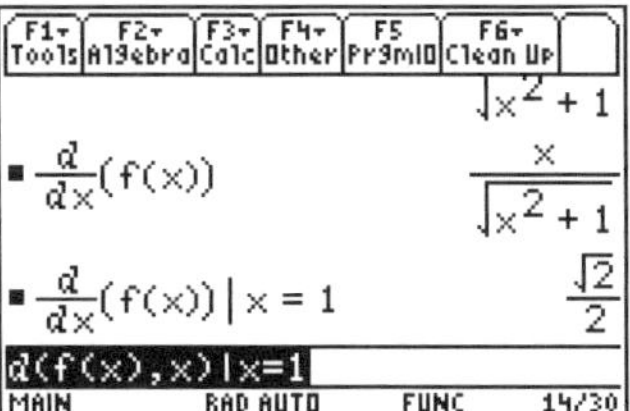

6. You also can evaluate the derivative at a particular point. The command uses the "with" operator that you used in Example 1 to estimate the slope at x=1. To compute $f'(1)$, press ▶ so the cursor is at the end of the entry line of your previous command. Then press [|] **X** [=] **1** [ENTER]. The exact slope at x=1 is displayed.

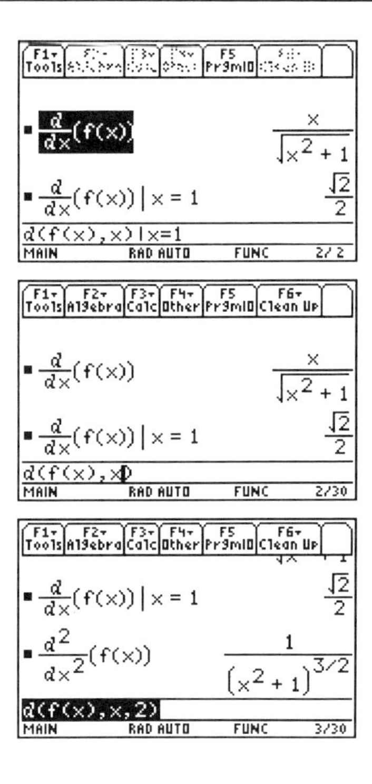

7. You can compute higher order derivatives by typing a value as the third argument of a differentiate command. To compute the second derivative, use the original derivative command (without x=1). You can move through the history area to recall that command. Move the insert cursor between the second x and the right parenthesis. Press [,] **2** [ENTER].

Example 3: Implicit differentiation

You can compute derivatives for implicit functions in several ways.

Given $x^2 + xy - y^2 + 2x - 3y - 12$, compute $\frac{dy}{dx}$ and evaluate it when x=4.

Solution

Use Implicit Function Theorem from multivariable calculus to compute the derivative in terms of x and y. Solve for the y-coordinates when x=4 and substitute. Alternately, enter y as an (unknown) function of x and compute the derivative directly. Finally, compute y as an explicit function of x, compute the derivative in terms of x only, and substitute.

Using multivariable calculus

Implicit Function Theorem states that if $z = f(x,y)$,

$$\text{then } \frac{dy}{dx} = \frac{-\partial z/\partial x}{\partial z/\partial y}.$$

Consult a multivariable calculus text for more details on this theorem.

1. First, store the expression to z.

   X [^] 2 [+] X [×] Y [−] Y [^] 2 [+] 2 X [−] 3 Y [−] 12 [STO▸] Z [ENTER]

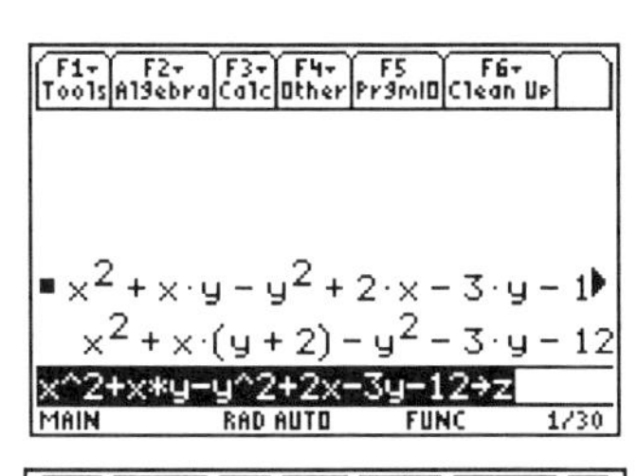

2. Apply Implicit Function Theorem and compute the result for implicit differentiation.

   [(-)] [2nd] [d] Z [,] X [)] [÷] [2nd] [d] Z [,] Y [)] [ENTER]

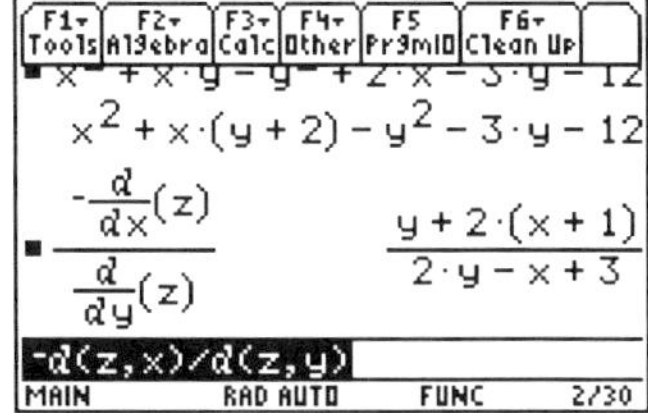

3. Solve for the y-coordinates when x=4.

   [F2] 1:solve(Z [=] 0 [,] Y [)] [|] X [=] 4 [ENTER]

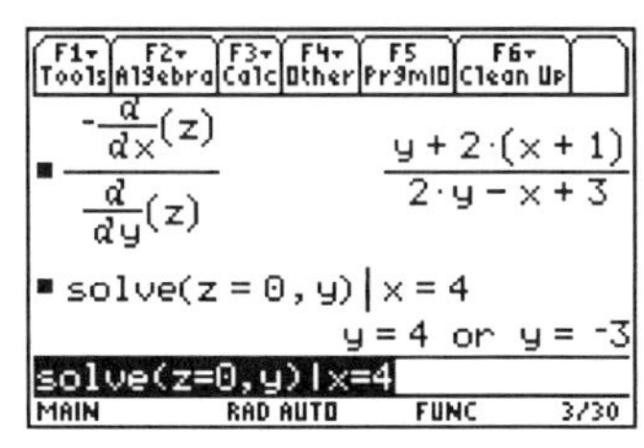

4. To compute the slope at (4,4), highlight the slope formula in the history area and press [ENTER]. Use the "with" operator and substitute for x and y to compute and display the slope (4,4).

   [|] X [=] 4 [CATALOG] and Y [=] 4 [ENTER]

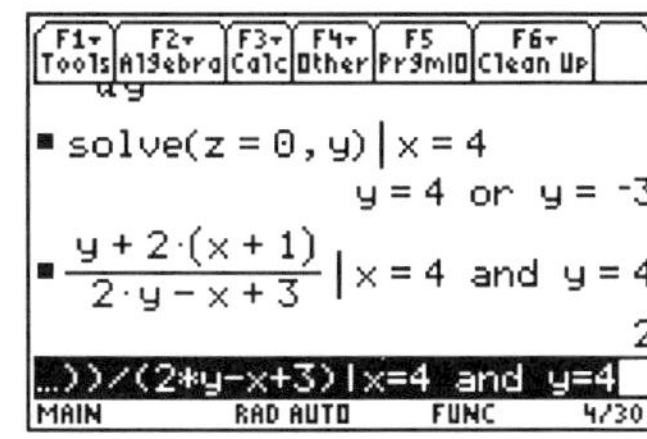

5. To compute the slope at the other point, press [▸] [←] [(-)] 3 [ENTER] to change the y-coordinate to -3.

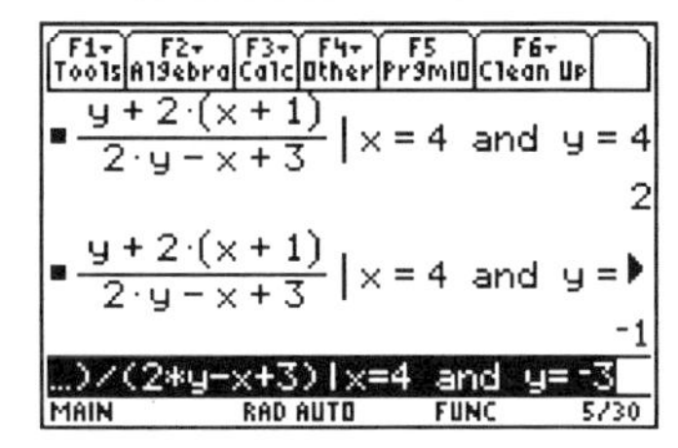

Computing the derivative directly

You may wish to compute the derivative without use of a multivariable calculus theorem. Two important details are: 1) it is necessary to write y as a function of x, rather than just the variable y and 2) some creativity is needed to get dy/dx alone for the final result.

1. Enter the differentiate command for d**(x^2 + x*y(x) − (y(x))^2 + 2x −3y(x) −12,x).**

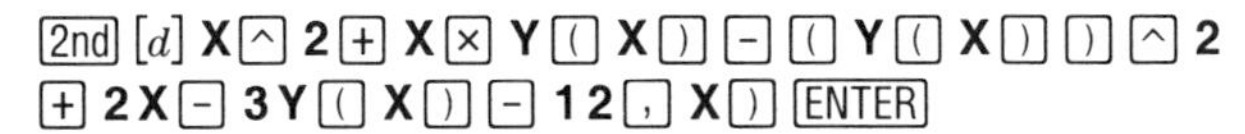

   [2nd] [d] X [^] 2 [+] X [×] Y [(] X [)] [−] [(] Y [(] X [)] [)] [^] 2 [+] 2 X [−] 3 Y [(] X [)] [−] 1 2 [,] X [)] [ENTER]

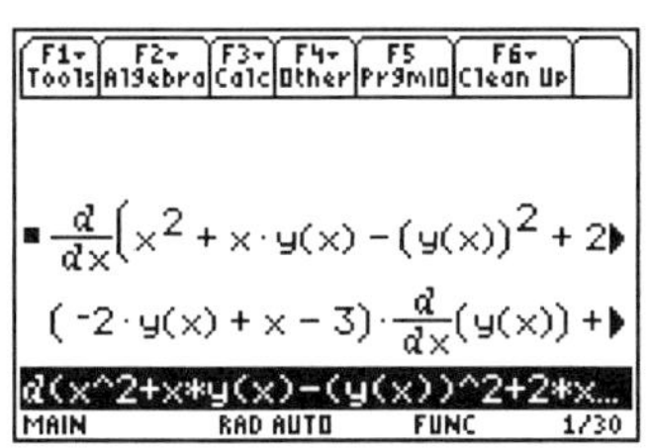

2. You now have a solution for the derivative. However, usually the solution is rearranged with $\frac{dy}{dx}$ isolated. The **solve(** command on the TI-89 will not accept the derivative as the variable to solve for, so you must substitute first. For instance, you might use $dydx$ (the four letters, not a built-in symbol).

   Press ⏶ [ENTER] to paste the previous to the entry line, and complete the expression.

   [|] [2nd] [*d*] **Y** [(] **X** [)] [,] **X** [)] [=] **DYDX** [ENTER]

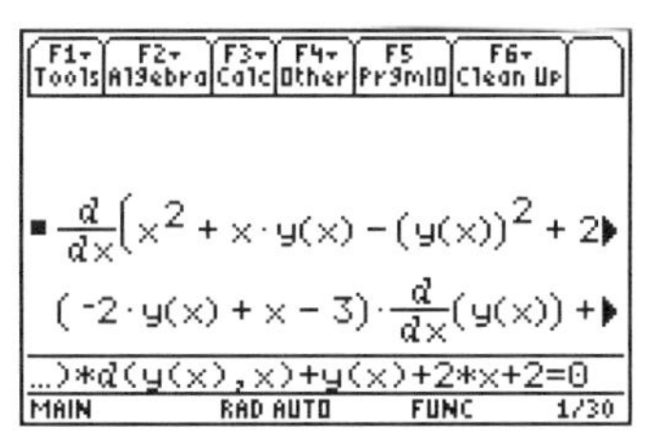

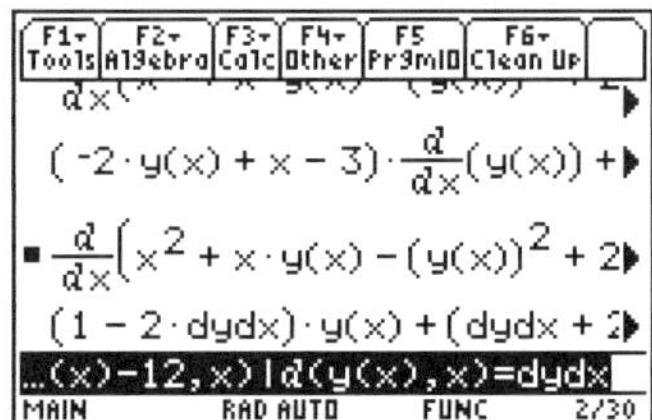

3. To solve, press [CLEAR] to clear the entry line. Then enter the command:

   [F2] **1:solve(** ⏶ [ENTER] [=] **0** [,] **DYDX** [)] [ENTER]

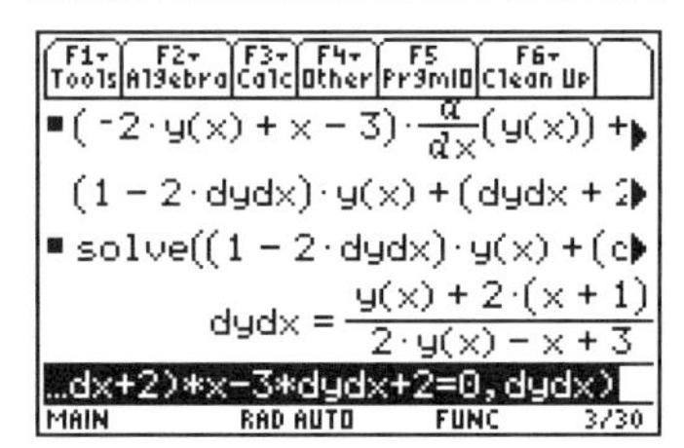

4. To evaluate the derivative at the point (4,4), edit the previous result.

   [CLEAR] ⏶ [ENTER] [|] **X** [=] **4** [CATALOG] **and Y** [(] **X** [)] [=] **4** [ENTER]

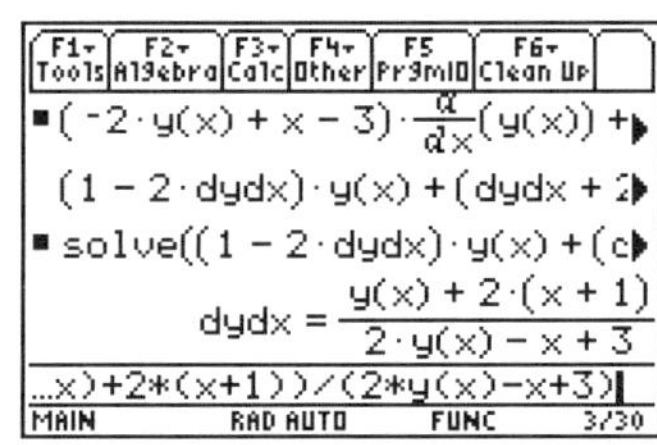

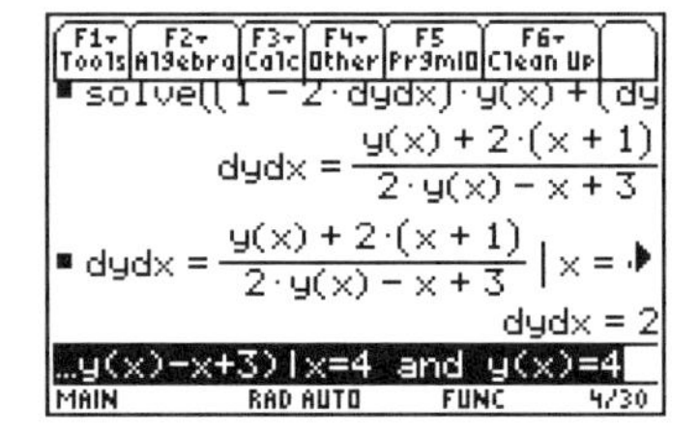

Solving for the function y(x)

In some, but not all implicit differentiation problems it is possible to find y as an explicit function of x. In this example, you can see that the relation is quadratic in y and you can solve for y.

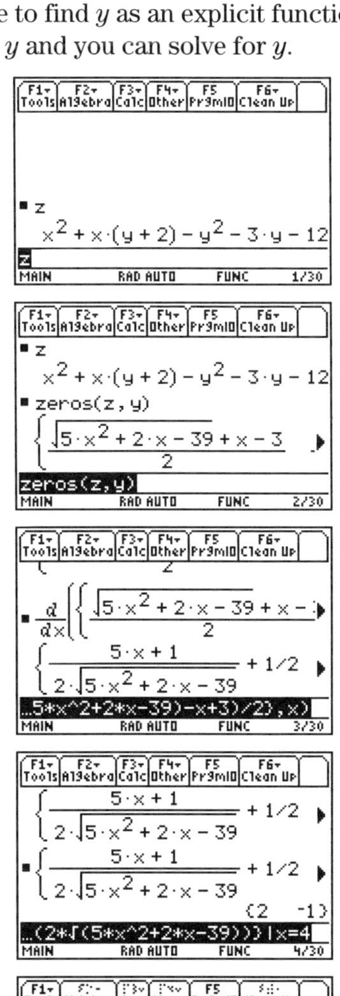

1. To solve for y in terms of x, first check that z is still defined. You can do this by pressing **Z** [ENTER].

2. Use **zeros(** to find the values of y that make z equal to zero.

   [F2] **4:zeros(Z** [,] **Y** [)] [ENTER]

3. To differentiate, press [2nd] [d] and then ⊖ [ENTER] to paste the previous result. Now press [,] **X** [)] [ENTER].

4. Evaluate the derivative when x=4.

   [CLEAR] ⊖ [ENTER] [|] **X** [=] **4** [ENTER]

5. You also can plot the implicit function and compute the value of the derivative at x=4 from the graph screen. For this approach, you can paste the result you have from the **zeros(** command to the Y= Editor. Press [CLEAR] and then press ⊖ five times to highlight the result of the **zeros(** command.

   Press [ENTER] to paste to the command line, [2nd] ⊙ to move the cursor to the beginning of the entry line, and ⊙ once to move the cursor between the { and the (.

Now press and hold [↑] while pressing ⊳ to highlight the first expression all the way to, but not including, the comma. Press [F1] **Tools** and select **5:Copy** to copy the expression.

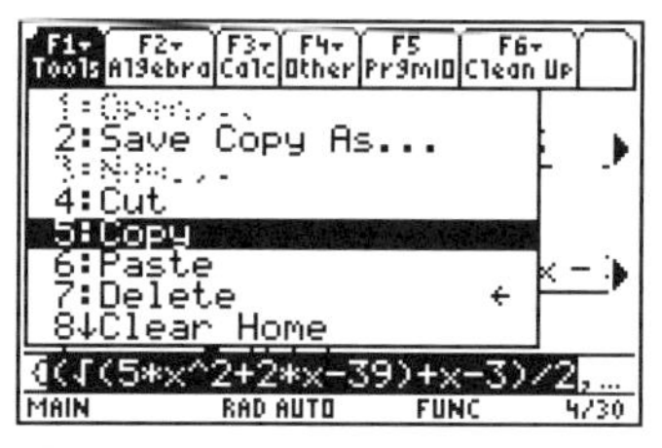

6. Press [◆] [Y=], check that the cursor is on $y1$, clear any function you may have in $y1$. Press [F1] **Tools,** and select **6:Paste**. The expression is pasted to the entry line. Press [ENTER] to paste the expression in $y1$.

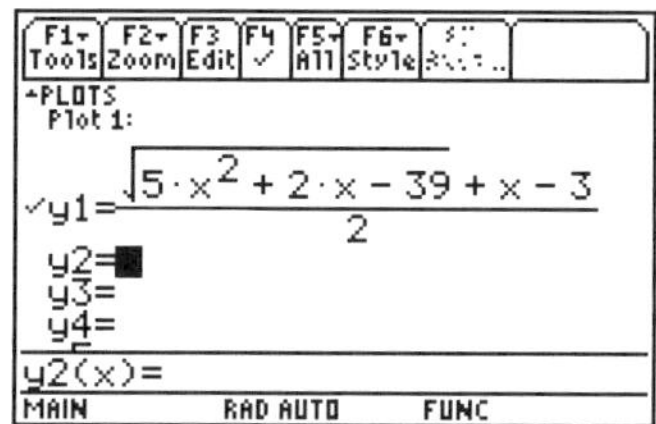

7. Press [HOME]. The previous command should still be on the entry line. Press ⊳ to move to the right of the comma. Press and hold [↑] while pressing ⊳ to highlight the second expression from the negative sign through the 2, that is, almost at the end but not including the last brace.

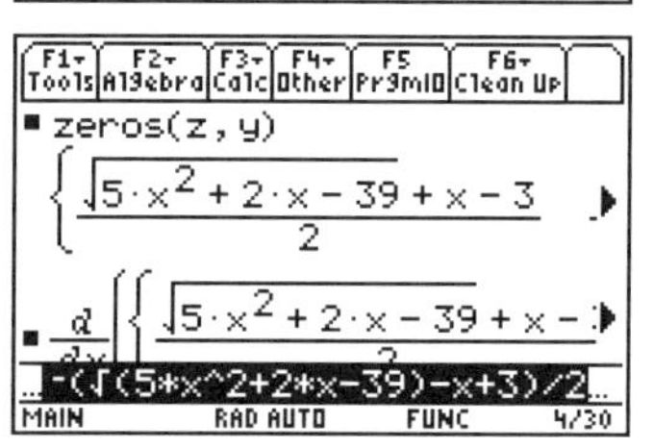

8. Copy and paste this expression to $y2$ in the same way you did with $y1$.

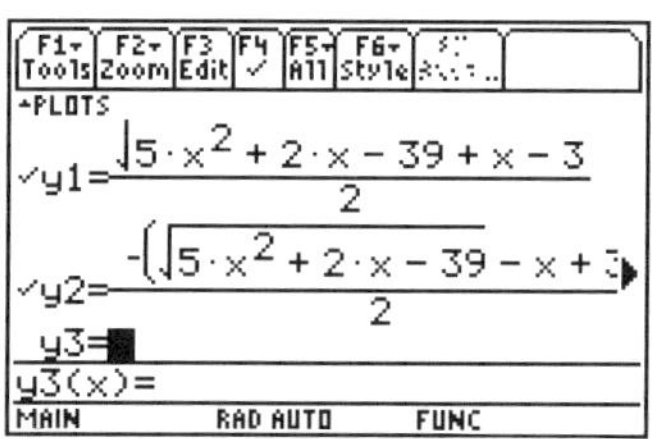

9. To set the standard Window variable values, press [F2] **Zoom** and select **6:ZoomStd**. The graph of the implicit relation is displayed.

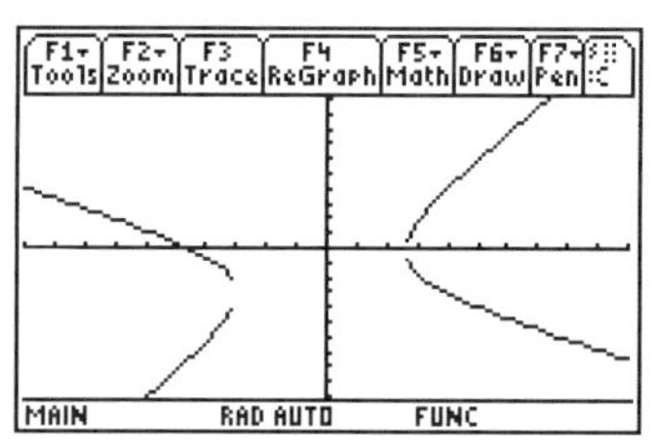

10. To draw a tangent line and display the equation, press [F5] **Math** and select **A:Tangent**. Enter **4** and press [ENTER].

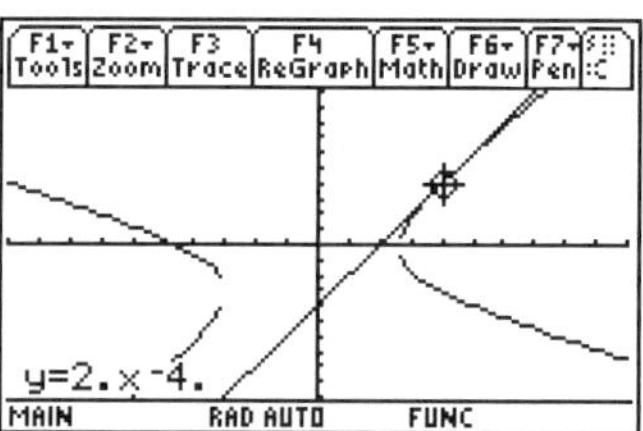

11. To draw the other tangent line and display the equation, repeat the tangent command and press ⊽ to move to the graph $y2$. Enter **4** and press [ENTER].

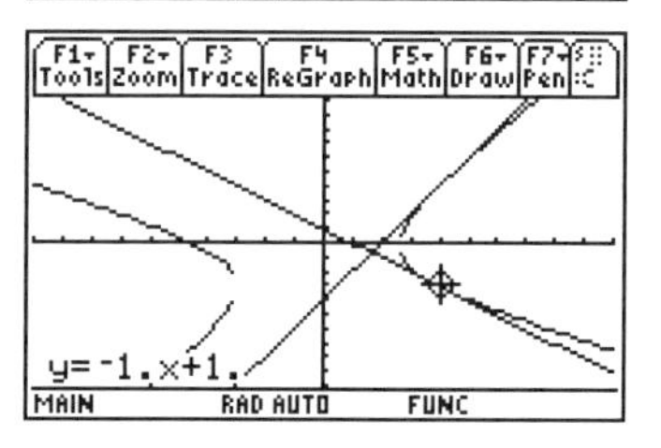

Example 4: Symbolic differentiation

It is important to remember that you have a calculator with a symbolic algebra system. Derivatives can be computed that help investigate rules of differentiation, such as the product rule and the chain rule.

Compute the derivatives for $\cos(u(x))$, and $(u(x))^n$, and $u(x) \cdot v(x)$.

Solution

Use the differentiate command with *u(x)* and *v(x)* as undefined functions.

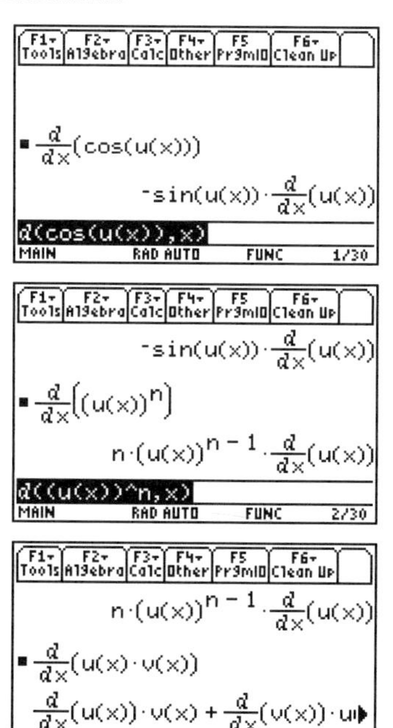

1. From the Home screen, press [2nd] [F6] **Clean Up** and select **1:clear a-z** to clear the variable name.
2. Differentiate $\cos(u(x))$ with respect to x.

   [2nd] [d] [2nd] [COS] **U** [(] **X** [)] [)] [,] **X** [)] [ENTER]

3. Differentiate $(u(x))^n$.

   [2nd] [d] [(] **U** [(] **X** [)] [)] [^] **N** [,] **X** [)] [ENTER]

4. Differentiate $u(x) \cdot v(x)$ in the same way.

   [2nd] [d] **U** [(] **X** [)] [×] **V** [(] **X** [)] [,] **X** [)] [ENTER]

Exercises

For exercises 1 to 8, use $f(x) = \sqrt{x^2 + 4x}$.

1. Compute the average rate of change from 2 to 2.001 and from 2 to 2.0001.
2. Compute the numerical derivative at x=2, using h=.001 and .0001.
3. Compute the limit of the average rate of change from 2 to 2+h as h approaches zero.
4. Compute the limit of the average rate of change from x to x+h as h approaches zero.
5. Compute $f'(x)$.
6. Compute $f'(2)$.
7. Compute the equation of the tangent line at x=2.
8. Compute $f''(x)$.

For exercises 9 to 12, use $3x^2 - xy + y^2 - 4x + 5y = 10$.

9. Compute $\frac{dy}{dx}$.
10. Compute y when x= ⁻2.
11. Compute the slope of the curve at all points where x= ⁻2.
12. Compute the equations for the tangent lines when x= ⁻2.
13. Compute the derivative for $f(x) = \frac{u(x)}{v(x)}$.

Chapter 3

Applications of the Derivative

In this chapter, you will explore two common applications of the derivative: optimization and related rate.

Example 1: Designing a cylinder

Many optimization problems involve volumes and surface areas. This example shows how to solve a classic cylinder problem with the TI-89.

A right circular cylinder with a top has a volume of 355 ml. Determine the dimensions of the cylinder with minimal surface area.

Solution

Define the surface area as a function of the radius. Compute the first derivative, set it equal to zero, and determine the minimum point. You also can obtain the same result on a graph with the minimization commands.

Solving numerically

1. Press [2nd] [F6] **Clean Up** and select **2:NewProb** to clear variables and set other defaults.
2. Store an expression for the volume to the variable v.

   [2nd] [π] [×] **R** [^] **2** [×] **H** [STO▸] **V** [ENTER].

   Store an expression for the surface area to the variable sa in a similar manner.

   2 [2nd] [π] [×] **R** [^] **2** [+] **2** [2nd] [π] [×] **R** [×] **H** [STO▸] **SA** [ENTER]

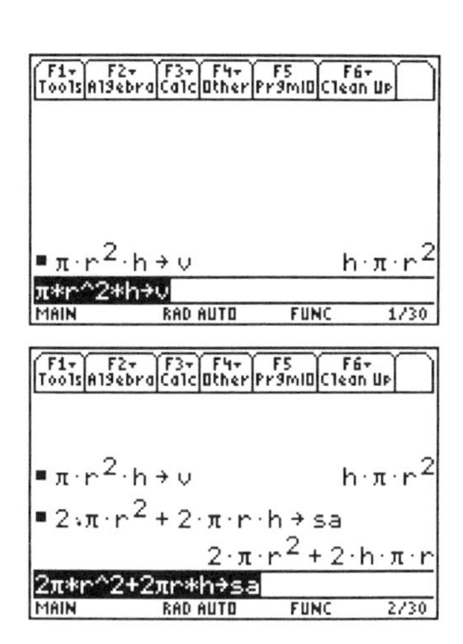

3. Since the volume is constant for this problem, you can use the **solve(** command to solve for h in terms of r.

   [F2] **1:solve(V** [=] **355** [,] **H** [)] [ENTER]

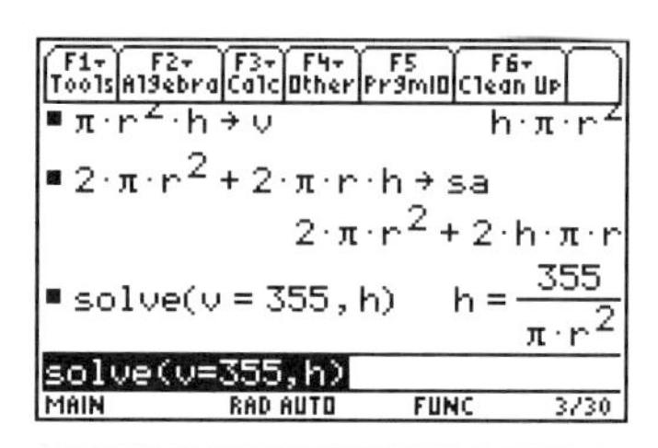

4. Substitute the result from step 3 into the surface area formula to express the surface area as a function of r only.

   SA [|] ⊖ [ENTER] [ENTER]

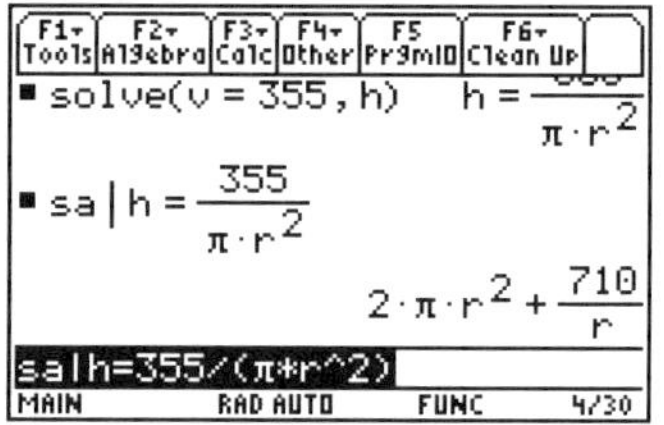

5. Compute and store the derivative of your surface area function.

   [2nd] [d] ⊖ [ENTER] [,] **R** [)] [STO▸] **DSA** [ENTER]

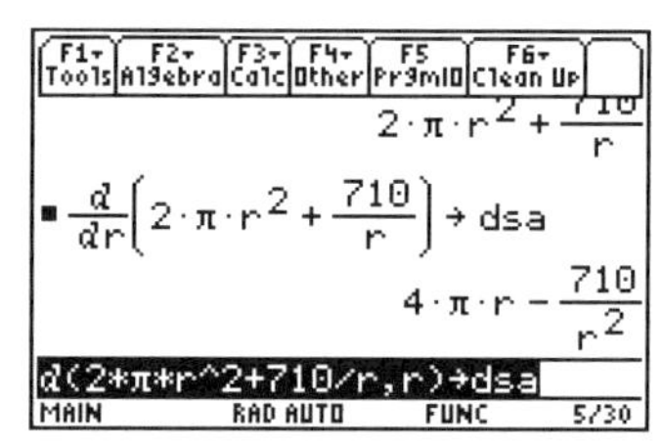

6. Use the **solve(** command to find the value of r when the derivative is zero.

   [F2] **1:solve(** ⊖ [ENTER] [=] **0** [,] **R** [)] [ENTER]

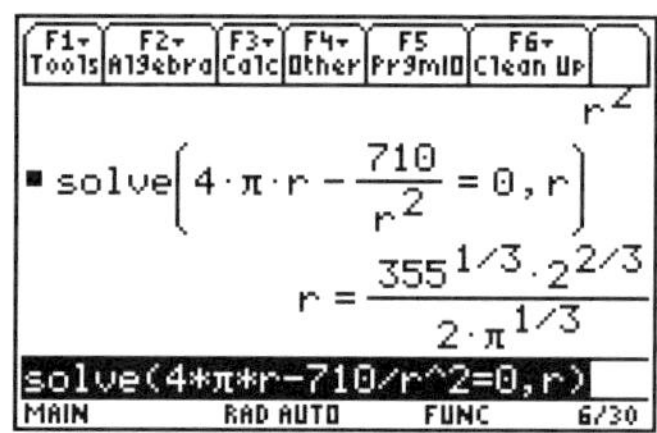

   To see a decimal estimate for the value, press [◆] [ENTER].

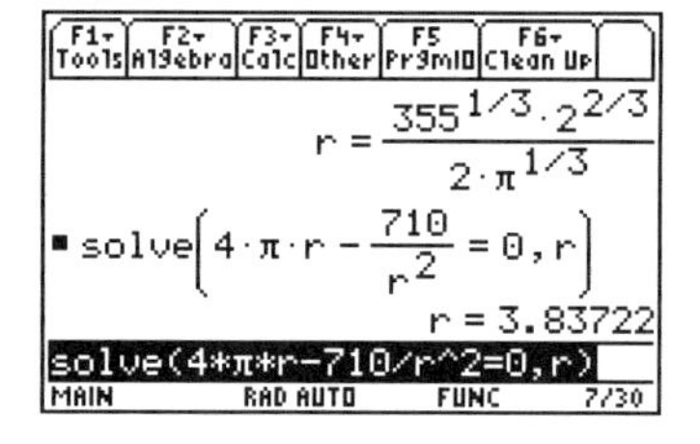

7. To test that the point is a minimum, you can compute the value of the derivative at a point on each side. First, test a point on the left.

   DSA [|] **R** [=] **3.8** [ENTER]

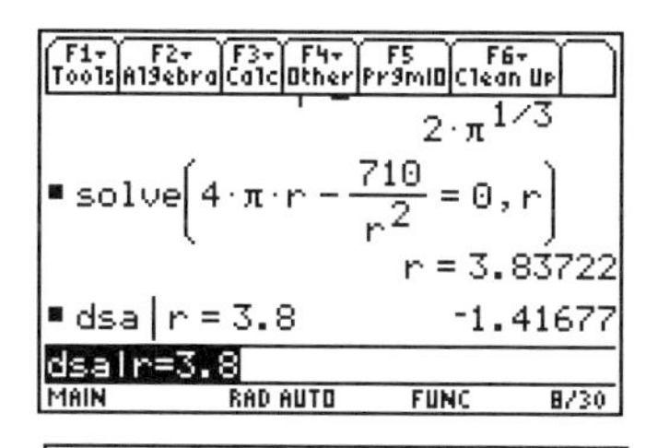

   Now test a point on the right.

   ⊙ [←] **9** [ENTER]

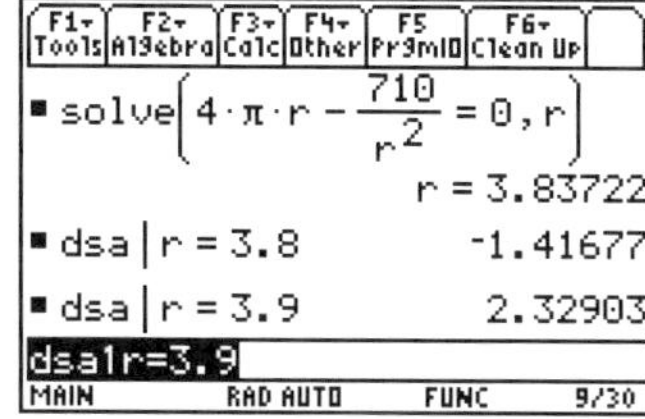

   The derivative values imply that the curve is falling and then rising, so your value is a minimum point.

8. Alternately, you can use the second derivative test to test that the point is a minimum. There are two ways to compute the second derivative. Since **dsa** is the derivative of **sa**, you can differentiate **dsa** and evaluate it at the value of r computed as the critical point in step 6.

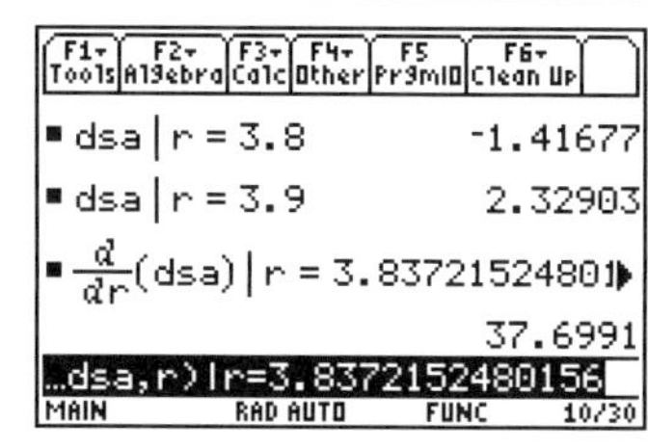

   You also can compute a second derivative of **sa** at the same critical point. Recall that a second derivative is computed when a 2 is used as the third argument of the differentiate command.

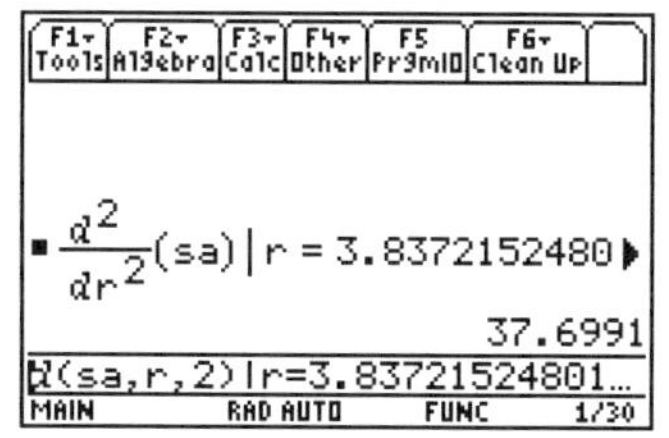

9. Compute the height of the cylinder using the value of r from step 6.

   355 [÷] [(] [2nd] [π] **R** [^] **2** [)] [|] ⊙(11 times) [ENTER] [ENTER]

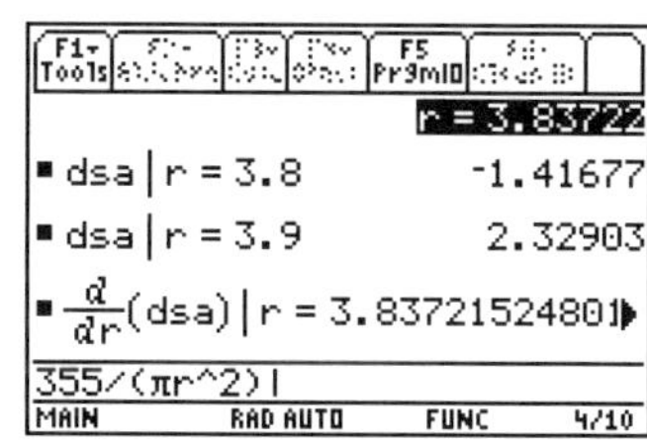

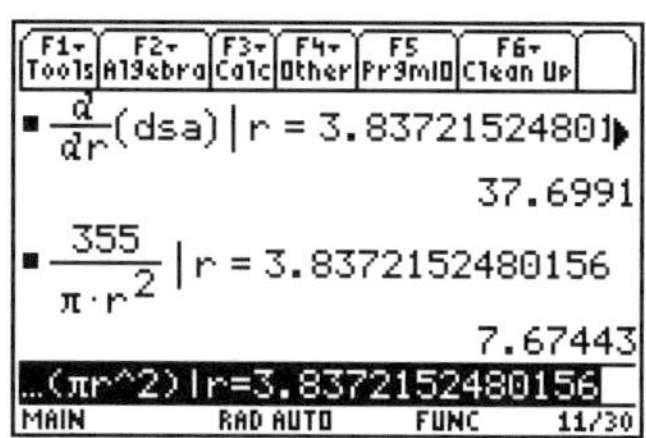

Solving graphically

1. Press [◆] [Y=] to display the Y= Editor. Press [CLEAR] as necessary to delete any functions. Define the function (the surface area formula from step 4 of *Solving Numerically*).

   2 [2nd] [π] **X** [^] **2** [+] **710** [÷] **X** [ENTER]

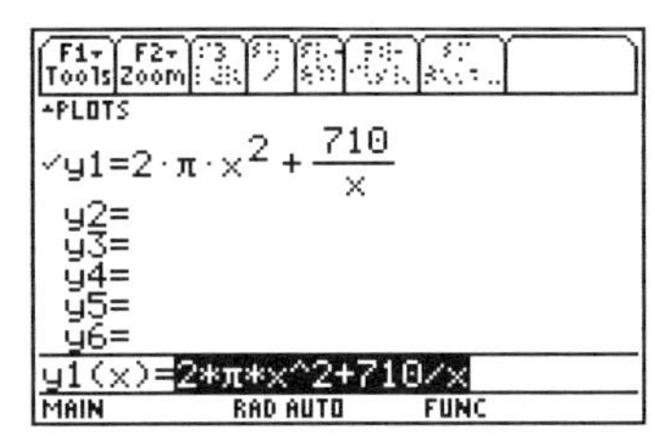

2. Press [◆] [WINDOW] and set the Window variable values as shown.

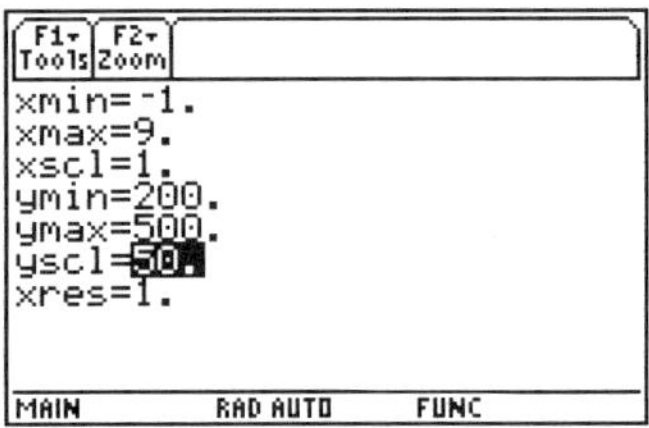

3. Press [◆] [GRAPH] to graph the function.

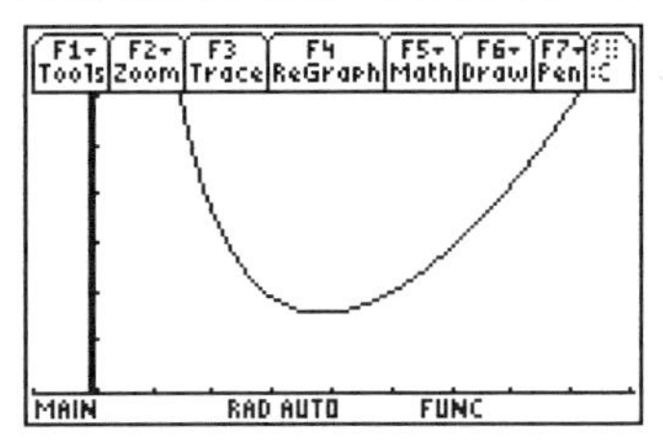

4. To compute the minimum point, press [F5] **Math** and select **3:Minimum**. Now use ⊙, or type a value to the left of the minimum point, and press [ENTER]. Press ⊙ or type a value for the right bound. The coordinates of the minimum point are displayed.

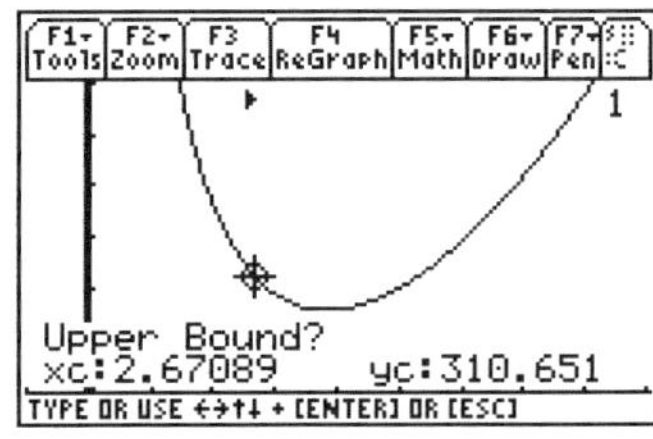

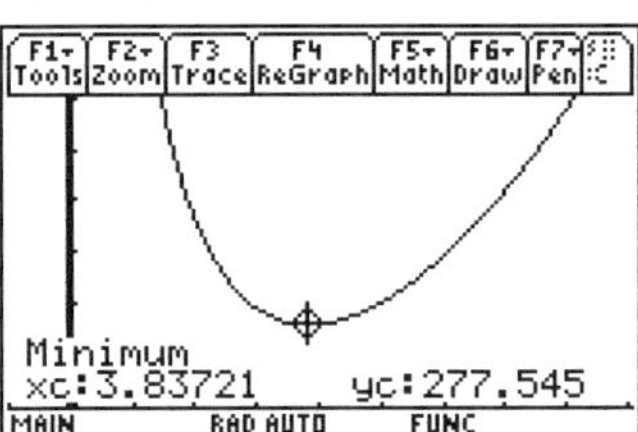

5. To compute a minimum on the Home screen without using derivatives, press [HOME] to return to the Home screen. Enter the command:

   [F3] **6:Min(Y1** [(] **X** [)] [,] **X** [)] [ENTER]

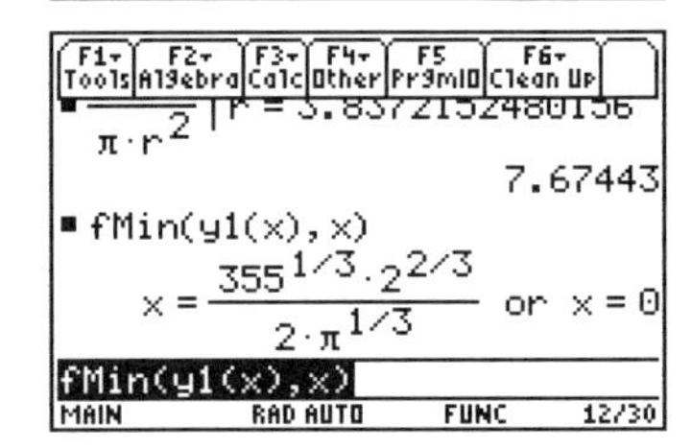

To see a decimal estimate for the value, press ⊖ [ENTER].

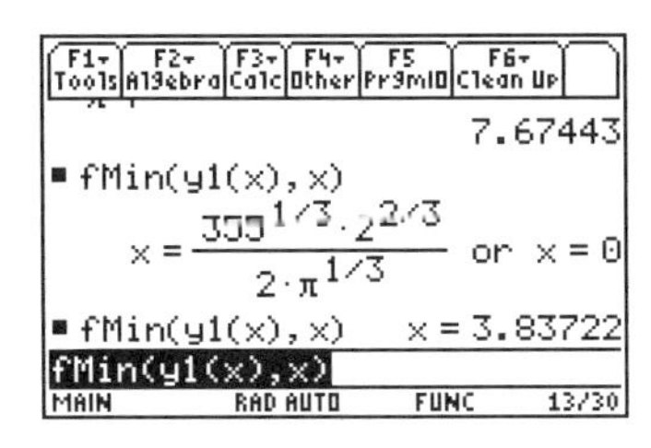

Example 2: Related rates

Many related rates examples investigate how the rates of change of two quantities are related. This classic sphere problem is done by considering both the volume and the radius as functions of time.

A spherical balloon is being inflated so that the radius is increasing at a steady rate of 2 cm/sec. Find the rate of change of the volume at any time t, and at the time when the radius is 8 cm.

1. Press [2nd] [F6] **Clean Up** and select **2:NewProb** to clear variables and set other defaults.

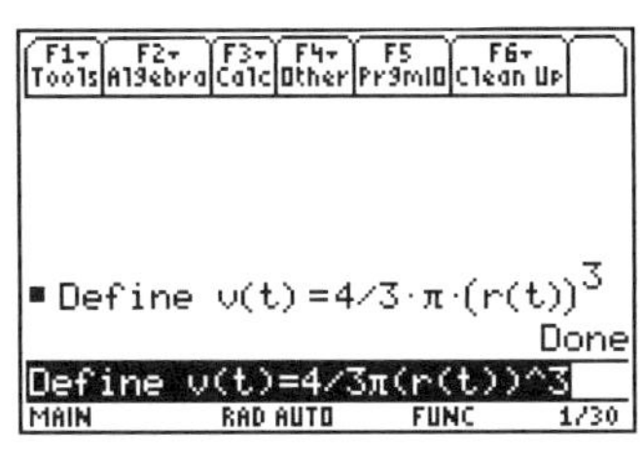

2. To define a volume function, press [F4] **Other** and select **1:Define**. Enter the volume function as shown. Note that the radius is actually a function $r(t)$ and that two sets of parentheses are needed to cube the radius.

3. Compute the derivative.

   [2nd] [d] **V** [(] **T** [)] [,] **T** [)] [ENTER]

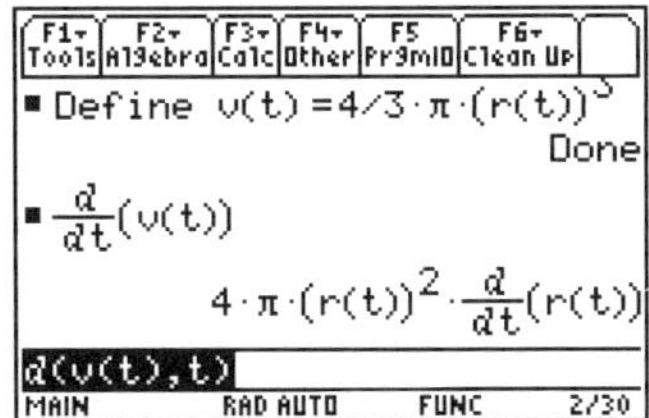

4. Substitute 2 for $\frac{dr}{dt}$ in the previous result to find the volume for t.

   ⊖ [ENTER] [|] [2nd] [d] **R** [(] **T** [)] [,] **T** [)] [=] **2** [ENTER]

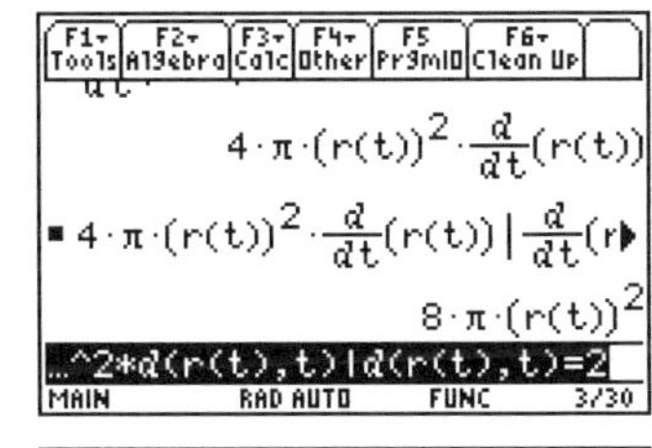

5. Use the previous result to find the volume when the radius is 8 cm.

   ⊖ [ENTER] [|] **R** [(] **T** [)] [=] **8** [ENTER]

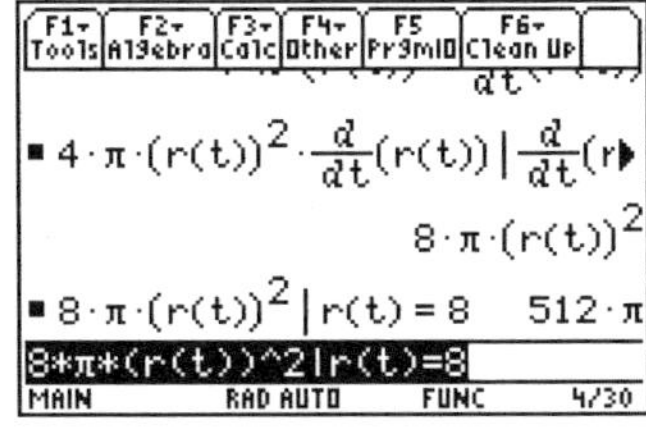

Exercises

Exercises 1 to 4 involve a right circular cylinder with no top that is constructed from 100 square cm of material.

1. Determine the volume as a function of the radius.
2. Compute the derivative of the volume function.
3. Use the derivative to determine the dimensions of the cylinder with maximum volume.
4. Compute the dimensions of the cylinder with maximum volume directly, without use of the derivative.

Exercises 5 and 6 involve a spherical iceball that is melting in such a way that the volume decreases at the rate of 6 cm^3 / sec.

5. Compute the derivative $\frac{dr}{dt}$ at any time t.
6. Compute the derivative $\frac{dr}{dt}$ at the time when the radius is 1 cm.

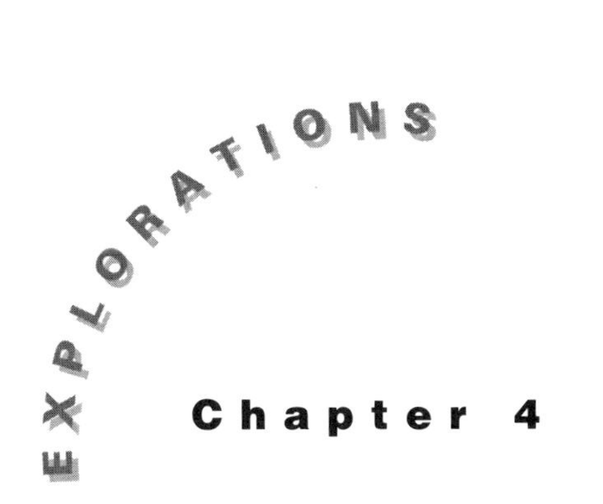

Chapter 4

Integration

In this chapter, you will explore indefinite and definite integrals.

Calculus courses describe many techniques of integration, such as integration by parts. The TI-89 can be used to verify the solutions to most problems.

Example 1: An indefinite integral

Integrate $\int x \cdot \cos(4x)dx$.

Solution

Use the integrate command (∫) on the Home screen.

1. Press [2nd] [F6] **Clean Up** and select **2:NewProb** to clear variables and set other defaults.

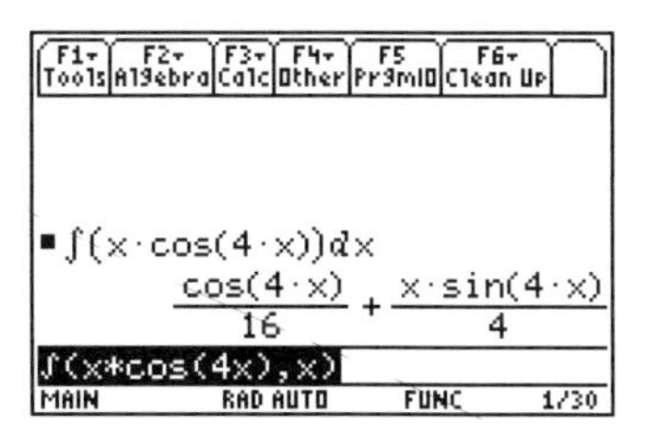

2. Enter the integral. The multiplication symbol after the first x is important; otherwise, the expression will be interpreted as an undefined function **xcos**.

   [2nd] [∫] **X** [×] [2nd] [COS] **4 X** [)] [,] **X** [)] [ENTER]

3. Of course, the complete solution is a family of curves, generally indicated by $+c$. In addition, the constant will be useful for further work, such as substituting an initial condition. Therefore, obtain a solution with the constant c.

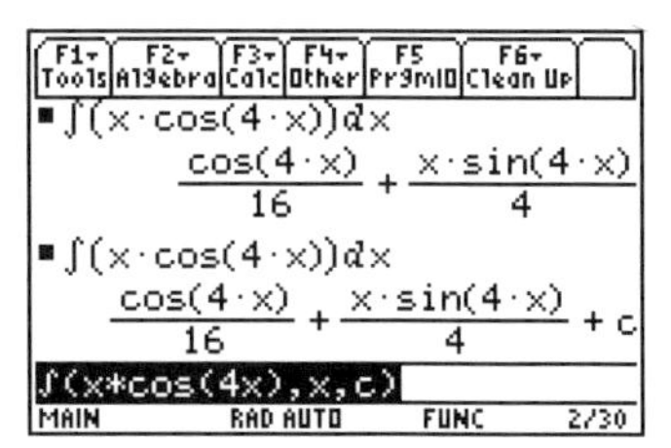

   [▶] [◀] [,] **C** [ENTER]

   Note that choices such as *c1* and *c2* cannot be used as they are reserved for the columns of the Data Matrix Editor.

Example 2: A definite integral

You can evaluate definite integrals with the TI-89, usually with an exact or approximate solution.

Evaluate

$$\int_{-2}^{3} (\frac{1}{8}x^3 - 2x)dx$$

Also, compute the area under $\frac{1}{8}x^3 - 2x$ on [-2,3].

Solution

Use the integrate command (∫) on the Home screen and the Graph screen.

1. Press [2nd] [F6] **Clean Up** and select **2:NewProb** to clear variables and set other defaults.

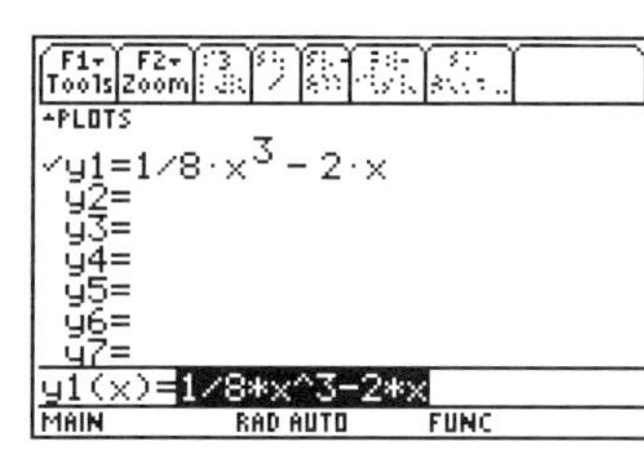

2. Press [♦] [Y=]. Clear any functions in the Y= Editor. With the cursor on $y1$, type the function and press [ENTER].

3. Press [F2] **Zoom** and select **4:ZoomDec** to graph the function.

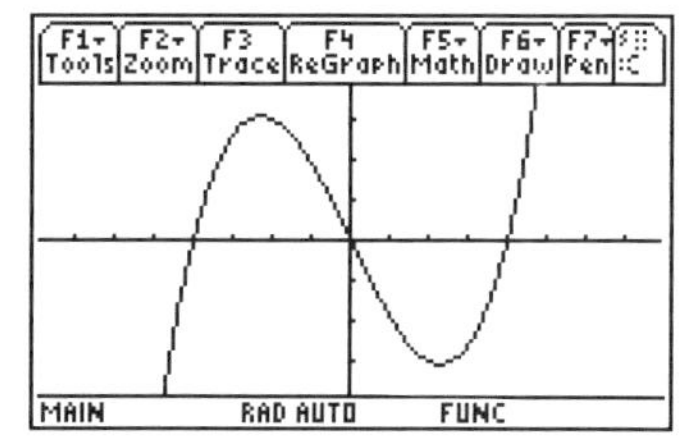

4. To evaluate the definite integral, press [F5] **Math** and select **7:∫ f(x) *d*x.** Type ⁻**2** as the lower limit and press [ENTER]. Type **3** as the upper limit and press [ENTER].

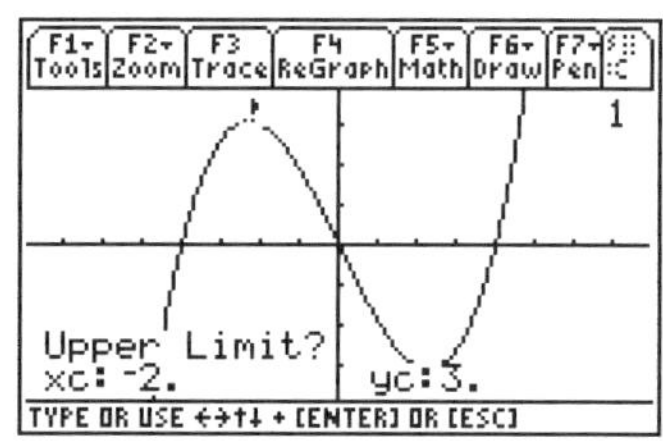

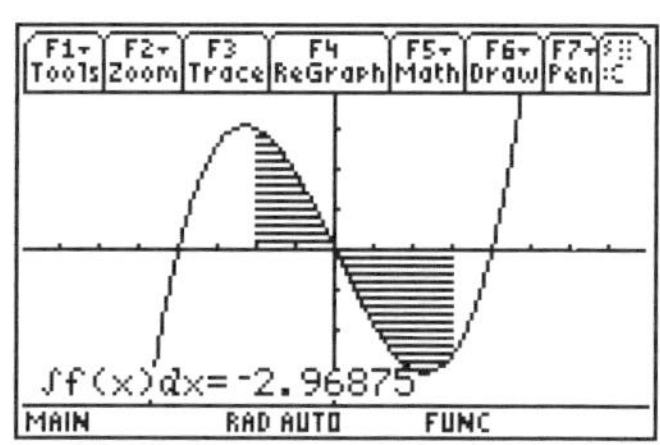

5. Press [HOME] to return to the Home screen and repeat the example.

   [2nd] [∫] **Y1** [(] **X** [)] [,] **X** [,] [(-)] **2** [,] **3** [)] [ENTER]

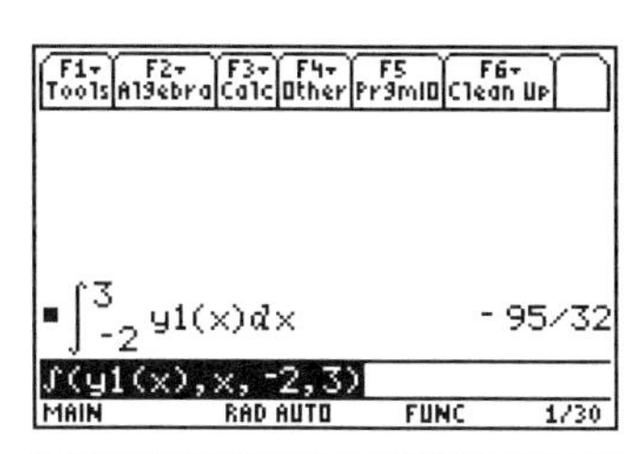

6. Since this function has a region below the x-axis on the interval [-2,3], the result for the area under the curve is not the same as the result for the definite integral computed above. There are several methods that can be used to compute the area. Since the curve is above the x-axis for [-2,0} and below the x-axis for [0,3], you can compute

$$\int_{-2}^{0} y1(x)dx - \int_{0}^{3} y1(x)dx$$

   ⓘ ⓘ [←] **0** ⓘ [−] [2nd] [∫] **Y1** [(] **X** [)] [,] **X** [,] **0** [,] **3** [)] [ENTER]

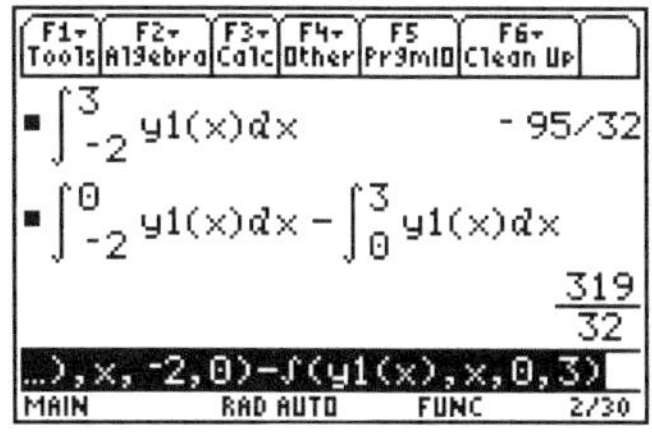

7. Alternately, you can compute this area by integrating the absolute value of $y1(x)$ on [-2,3] using **abs(**.

   [2nd] [∫] [CATALOG] **abs(Y1** [(] **X** [)] [)] [,] **X** [,] [(-)] **2** [,] **3** [)] [ENTER]

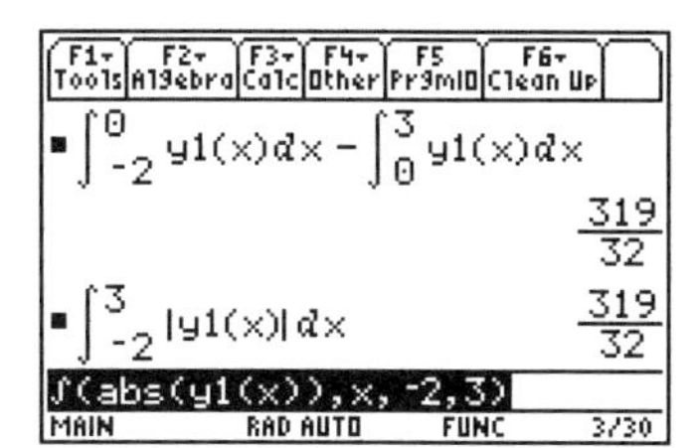

Example 3: An integral formula

Since the TI-89 has a symbolic algebra system, integrals with undeclared coefficients can be computed. The results resemble those found in tables of integrals. They are beneficial for generating formulas and pattern recognition.

$$\text{Integrate } \int \frac{1}{a^2 + (bx)^2} dx.$$

Solution

Use the integrate command (∫) on the Home screen.

1. Press [2nd] [F6] **Clean Up** and select **2:NewProb** to clear variables and set other defaults.

2. Enter the function. Recall that it is important to type **b*x** not **bx**. You can also use undeclared variables in the limits.

   [2nd] [∫] **1** [÷] [(] **A** [^] **2** [+] [(] **B** [×] **X** [)] [^] **2** [)] [,] **X** [)] [ENTER]

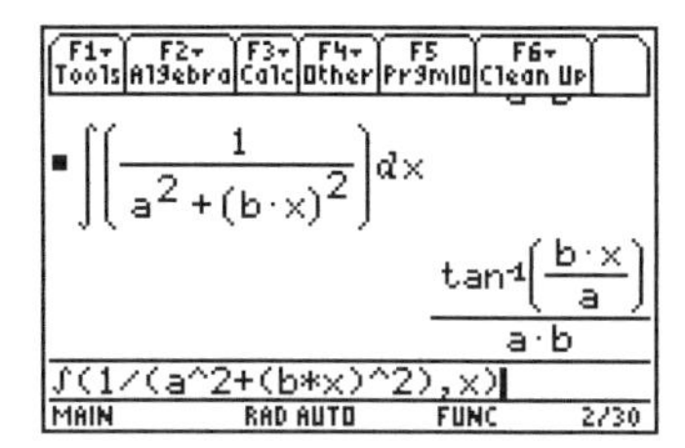

Example 4: Symbolic limits in a definite integral

Evaluate $\int_a^b \frac{1}{k+t}\,dt$

Solution

Use the integrate command (∫) on the Home screen.

1. Press [2nd] [F6] **Clean Up** and select **2:NewProb** to clear variables and set other defaults.

2. Enter the definite integral.

[2nd] [∫] 1 [÷] [(] K [+] T [)] [,] T [,] A [,] B [)] [ENTER]

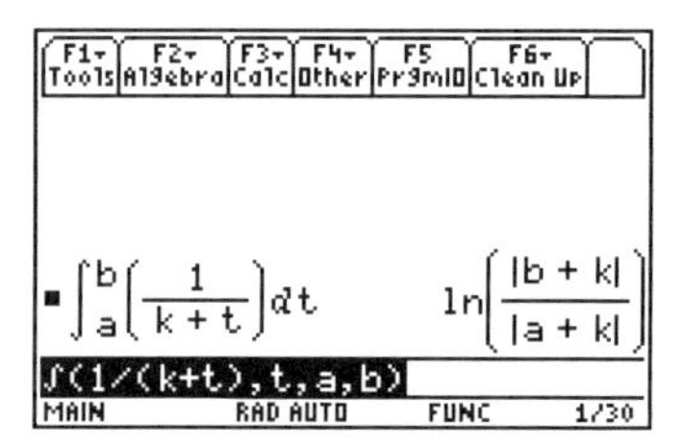

Exercises

Integrate each example.

1. $\int \frac{x}{1+4x^2}dx$

2. $\int a \cdot b^{k \cdot t} dt$

3. $\int_0^{\frac{\pi}{3}} \sin(x)\cos(x)dx$

4. $\int_p^q (m \cdot x + b)^n dx$

Chapter 5

Riemann Sums and the Fundamental Theorem of Calculus

In calculus you study two types of integrals: indefinite integrals and definite integrals. *Indefinite integrals* are used to find the antiderivative of a function. *Definite integrals* can be used to find the area bounded by a function and the x-axis. In the following examples, you will discover a remarkable connection between these two types of integrals called the Fundamental Theorem of Integral Calculus.

Example 1: The area under a parabola

Example 1 begins with a numerical method to approximate the area bounded by a curve and then uses an analytic method that gives the exact area. The numerical method includes detailed graphs and rather lengthy calculations, which are performed with a calculator program. This program is listed at the end of this chapter. To duplicate the graphs and calculations, you will need to enter this program using the Program Editor in the APPS menu before beginning Example 1. If you don't want to enter the program, you can still study the numerical method and then use your TI-89 when you get to the analytic method.

Find the area bounded by

$$f(x) = x^2,\ x = 0,\ x = 1 \text{ and } y = 0.$$

Solution

Before finding the area using either the numerical or analytic method, graph the function. Enter the function as $y1$ in the Y= Editor. Enter the Window variable values shown here. Then graph the function.

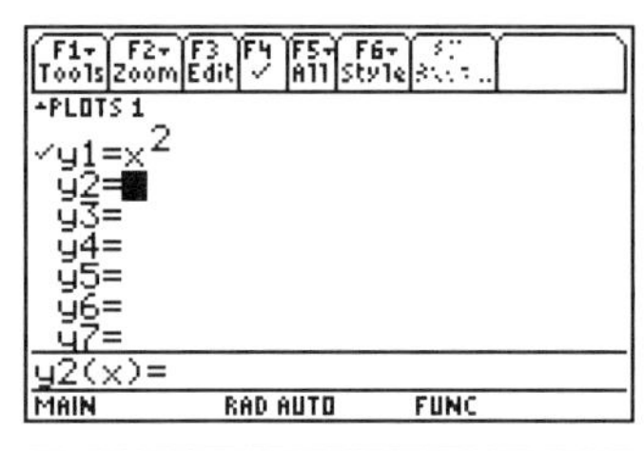

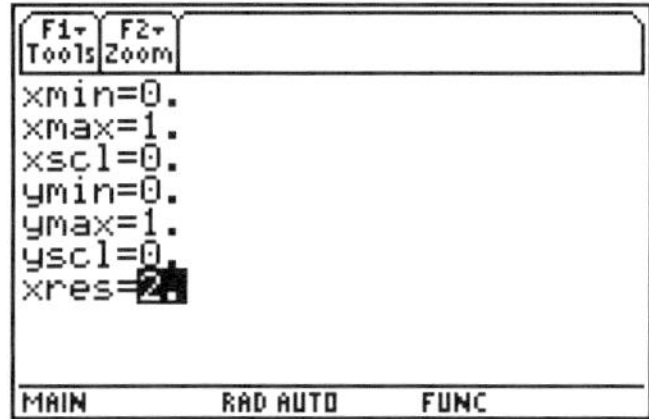

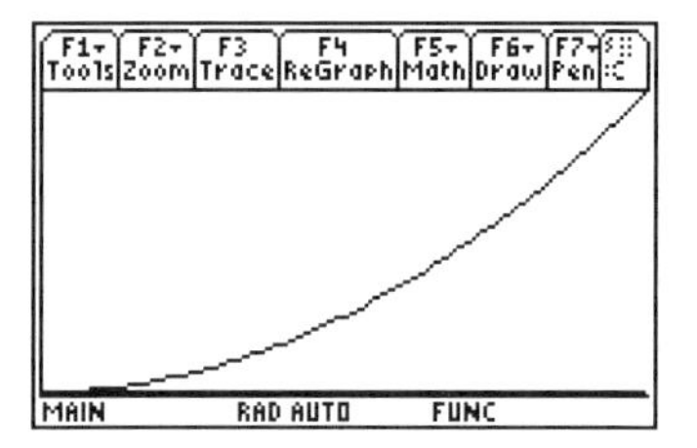

Numerical method

You can estimate the area under the curve with inscribed rectangles.

1. In the Program Editor, define the **area()** program listed at the end of this chapter.

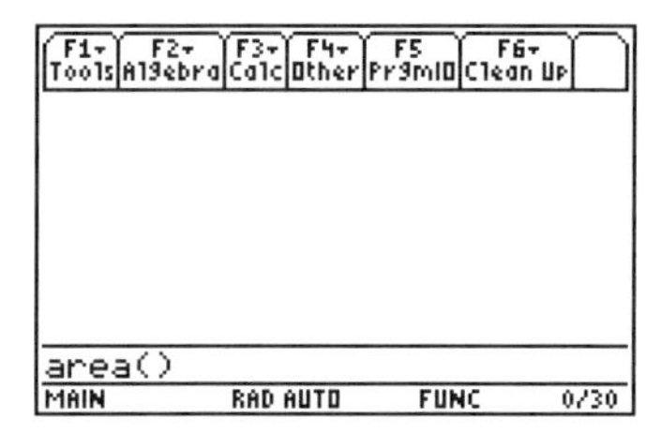

2. After you have entered the function in $y1$ and graphed it, return to the Home screen, type **area()**, and press [ENTER] to run the program.

3. The program prompts the user for the left and right boundaries of the area and for the number of rectangles. Enter **0** and **1** for the boundaries and **10** for the number of rectangles.

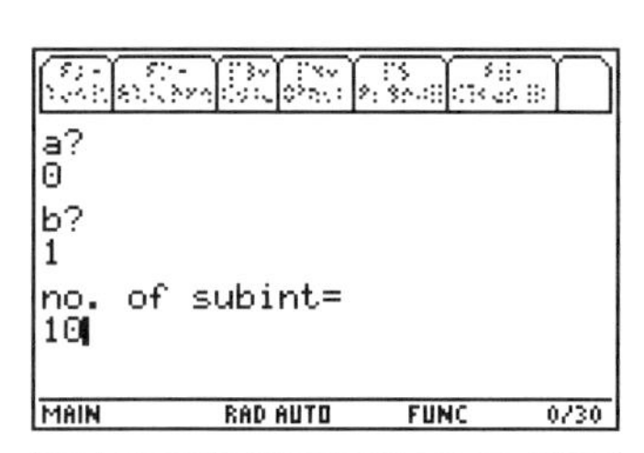

The program first inscribes rectangles with upper-left-hand corners on the curve.

$$f(x)=x^2$$

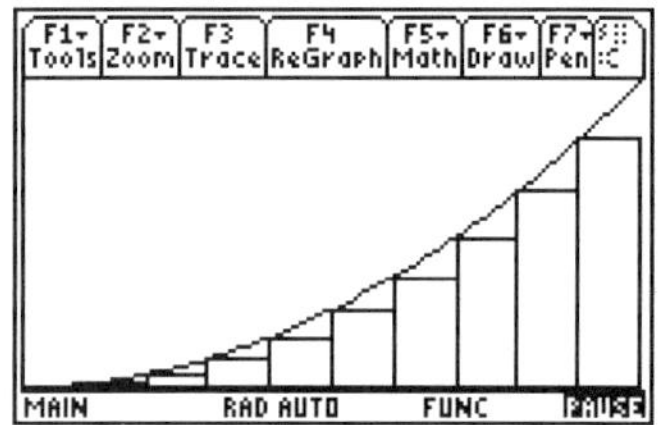

4. The sum of the areas of these left-hand rectangles approximates the area under the curve. This is called a left-hand rectangular approximation method, **lram**. Press [ENTER] to see the sum.

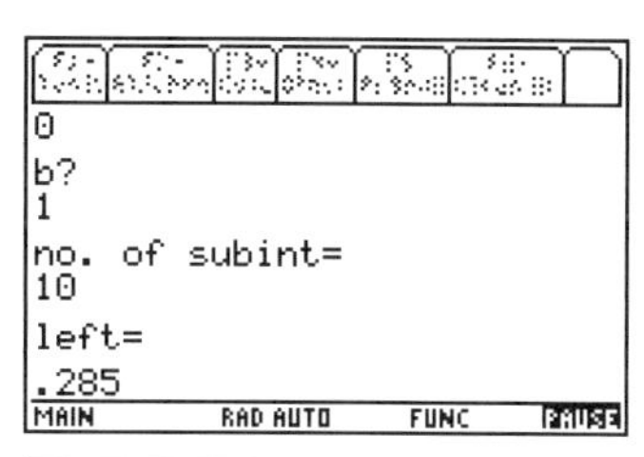

5. Next the program estimates the area with a right-hand rectangular approximation method, **rram.** The upper-right-hand corners of these rectangles lie on the curve. Press [ENTER] to see these rectangles, and then press [ENTER] again to see the sum of the areas of these rectangles.

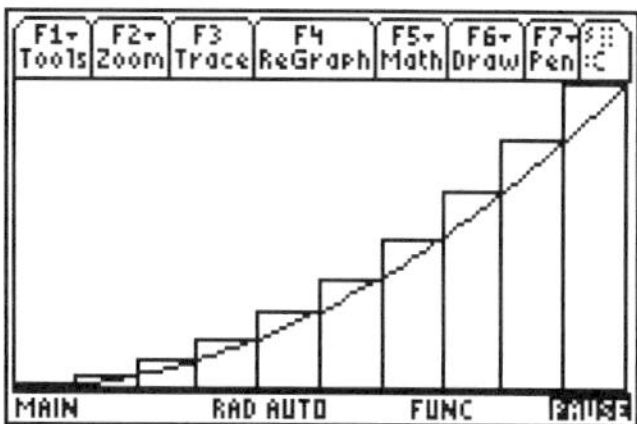

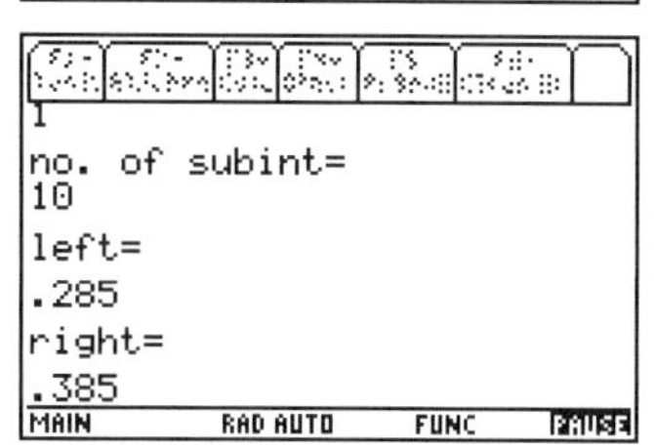

6. The program now uses rectangles with midpoints on the curve, called **mram**. Press [ENTER] to see these rectangles, and then press [ENTER] again to see the sum of their areas.

   The sums of the left, right, and midpoint rectangles are called *Riemann Sums*. Which of the three Riemann Sums do you think best approximates the area bounded by

$$f(x) = x^2,\ x = 0, \text{ and } y = 0?$$

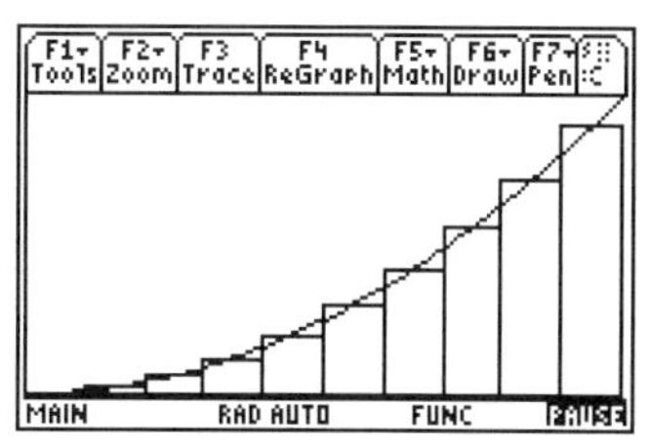

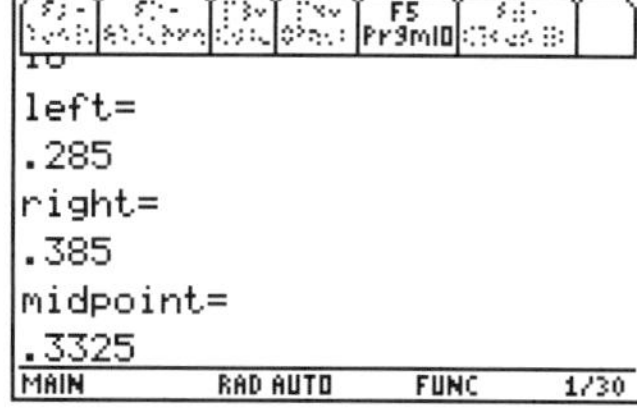

What do you think will happen if you increase the number of rectangles to 25?

Press [HOME] to return to the Home screen and run the **area()** program again with 25 subintervals.

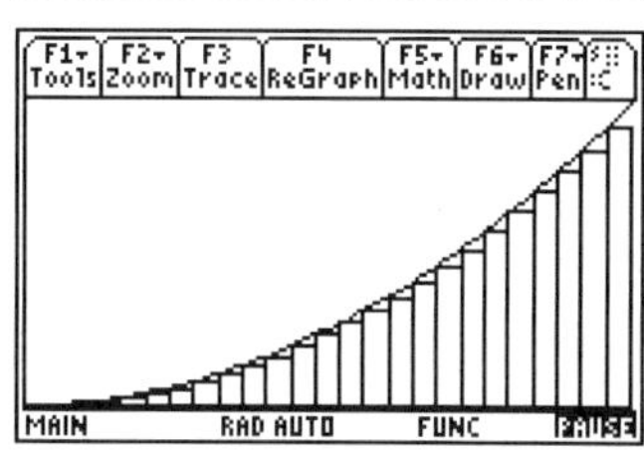

It appears that as the number of rectangles increases, they form a better fit to the area being estimated. Notice that the left, right, and midpoint estimates seem to be converging. From this, we estimate that the area is about .33.

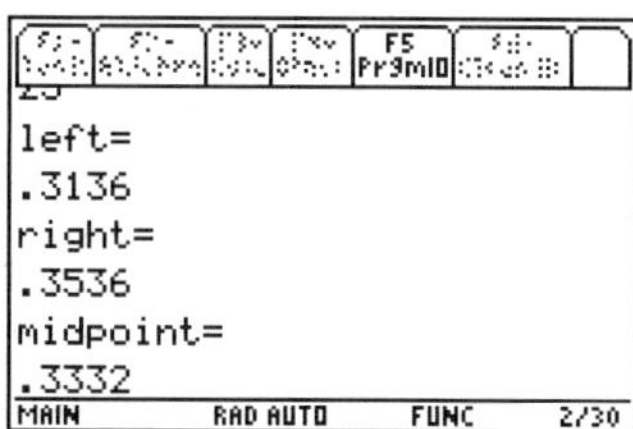

Analytic method

You can find the exact value to which the left, right, and midpoint Riemann Sums converge with an analytic approach. First, define TI-89 functions for each of the rectangle methods **lram**, **rram**, and **mram**. Then take their limit as the number of rectangles approaches infinity in order to find the value to which they converge.

Let's reexamine a picture of the right-hand rectangles in order to see how to define **rram** symbolically.

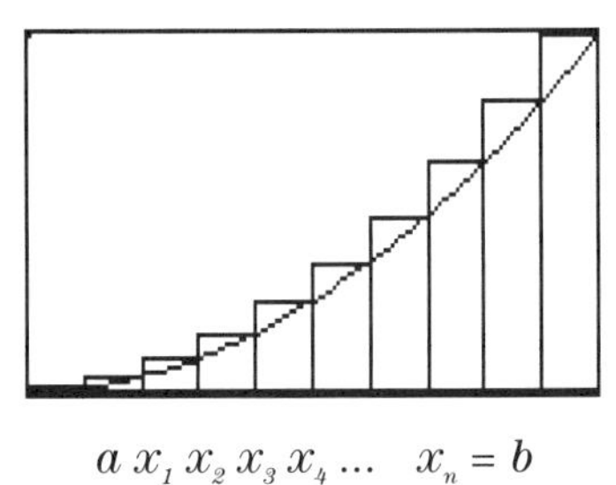

$a\ x_1\ x_2\ x_3\ x_4 \ldots\ x_n = b$

The area of each rectangle is given by the product of its height and width. If n is the number of rectangles and a and b are the left and right boundaries of the area, then the width of each rectangle is

$$h = \frac{b-a}{n}$$

The height of each rectangle is given by

$$f(x_1),\quad f(x_2),\quad f(x_3),\quad \ldots\quad f(x_n)$$

where

$$x_1 = a+h,\quad x_2 = a+2h,\quad x_3 = a+3h,\quad \ldots\quad x_k = a+k\cdot h$$

This leads to the following definition for **rram**:

$$rram = h \sum_{k=1}^{n} f(a+k\cdot h)$$

The definitions for **lram** and **mram** are similar in nature to **rram**. Study the TI-89 function definitions below, and try to understand how they work.

$f(x)=x^2$

lram$(a,b,n)=h*\Sigma(f(a+h*k),k,0,n\text{-}1)|h=(b\text{-}a)/n$

rram$(a,b,n)=h*\Sigma(f(a+h*k),k,1,n)|h=(b\text{-}a)/n$

mram$(a,b,n)=h*\Sigma(f(a+h*(.5+k)),k,0,n\text{-}1)|h=(b\text{-}a)/n$

Define these functions on your TI-89 and find the limit to which they converge.

1. Press [2nd] [F6] **Clean Up** and select **2:NewProb** to clear variables and set other defaults.
2. Use the **Define** command to define the functions shown above. After you define **lram**, you can edit the entry line to change **lram** slightly to produce **rram** and **mram** so that you don't have to type their entire definitions.

   [CATALOG] **Define F** [(] **X** [)] [=] **X** [^] **2** [ENTER]

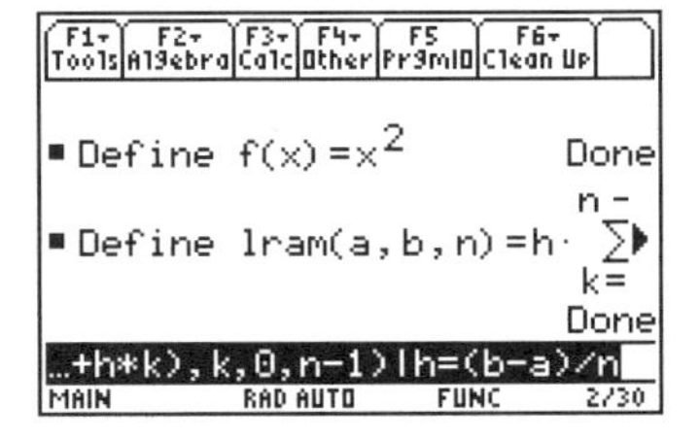

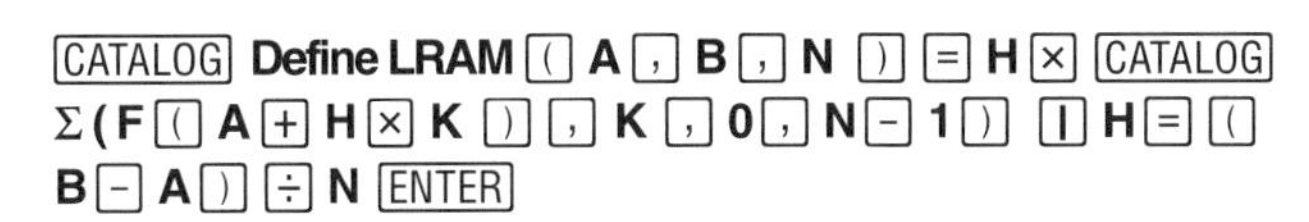

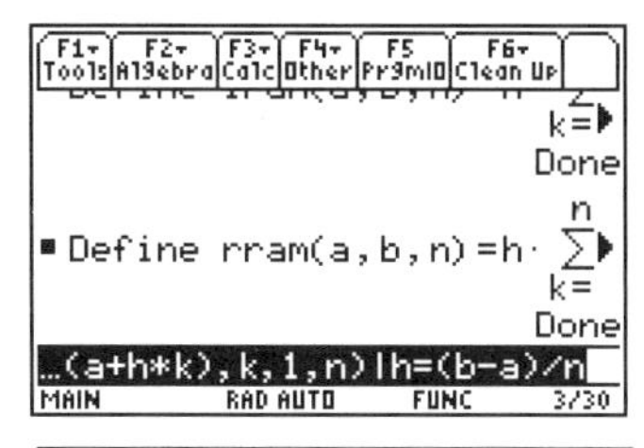

3. Set the Exact/Approx mode setting on Page 2 of the MODE dialog box to **APPROXIMATE**.

4. Check these functions by evaluating each one for $a = 0$, $b = 1$, and $n = 10$.

```
■ lram(0,1,10)        .285
■ rram(0,1,10)        .385
■ mram(0,1,10)        .3325
mram(0,1,10)
MAIN   RAD APPROX   FUNC   3/30
```

5. Set Exact/Approx= **EXACT**.

6. Take the limit of each summation function as the number of rectangles approaches infinity. The exact area is 1/3.

```
■ lim lram(0,1,n)     1/3
  n→∞
■ lim rram(0,1,n)     1/3
  n→∞
■ lim mram(0,1,n)     1/3
  n→∞
limit(mram(0,1,n),n,∞)
MAIN   RAD EXACT   FUNC   3/30
```

7. The definite integral $\int_0^1 f(x)dx$ also should equal the limit of the Riemann Sums.

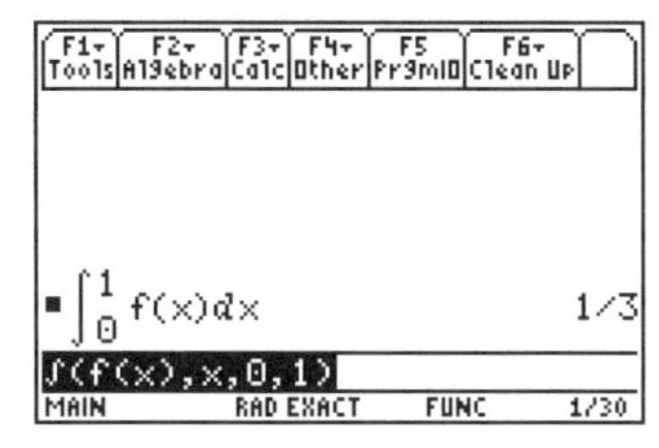

[2nd] [∫] F [(] X [)] [,] X [,] 0 [,] 1 [)] [ENTER]

8. Find the area for other values of b ($b=2$ and $b=3$). See if the definite integral gives the same result as the limits of the Riemann Sums.

```
■ lim lram(0,2,n)     8/3
  n→∞
■ lim rram(0,2,n)     8/3
  n→∞
■ lim mram(0,2,n)     8/3
  n→∞
limit(mram(0,2,n),n,∞)
MAIN   RAD EXACT   FUNC   3/30
```

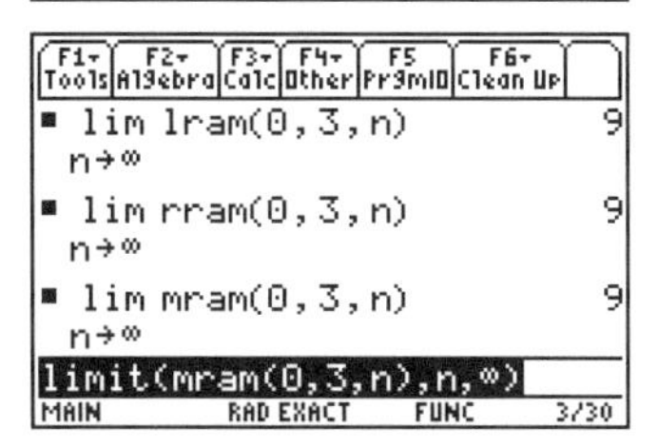

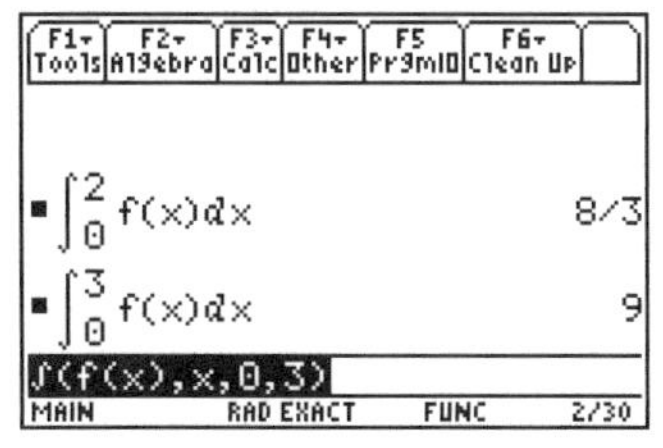

9. Predict the result for $a = 0$ and $b = x$. Verify your prediction on the TI-89.

   How is the area from $a = 0$ to $b = x$ related to $f(x)$? If you noticed the area function is the antiderivative of $f(x)$, you are on your way to discovering a relationship called the Fundamental Theorem of Calculus.

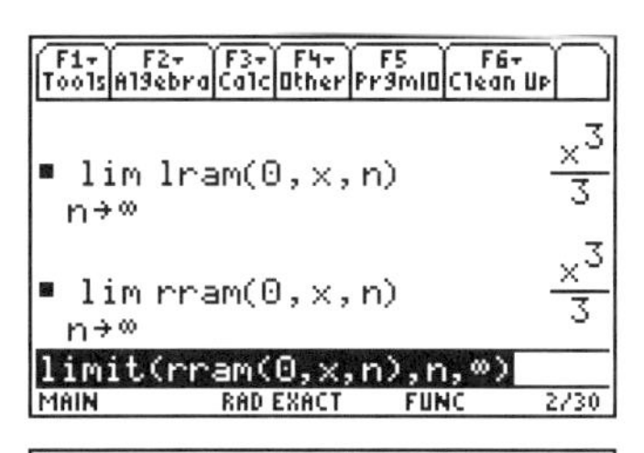

10. Use **lram** to find the area from a to b.

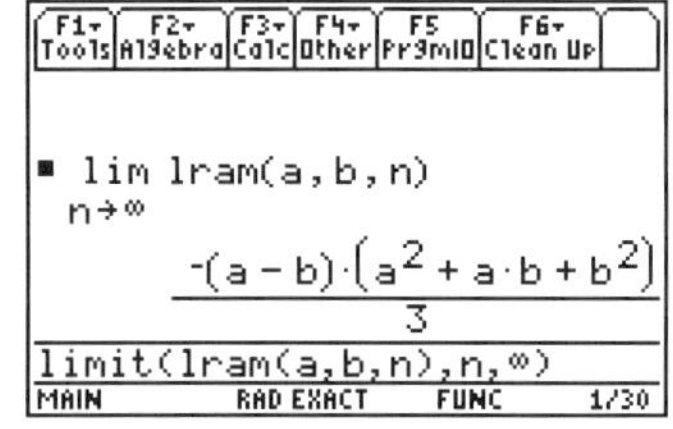

11. This answer is not in an easily recognizable form, so use the **expand(** command to expand the result. The result from step 10 is pasted from the history area to the entry line to save typing.

   [CATALOG] **expand(** ⊖ [ENTER] [)] [ENTER]

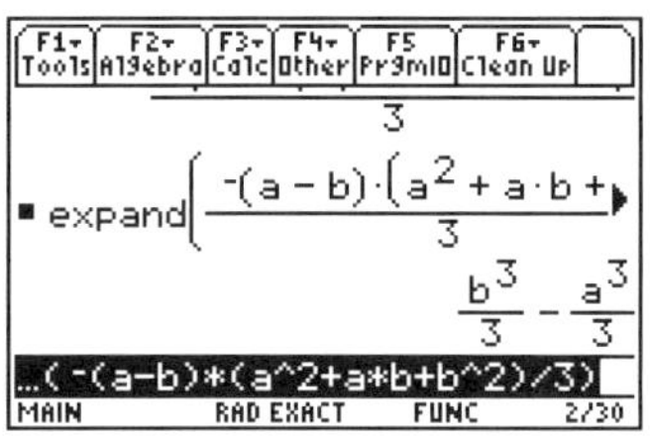

12. Predict $\int_a^b f(x)dx$.

   Verify your prediction on the TI-89.

   How is this answer related to $f(x)$?

   Did you notice there is an antiderivative of $f(x)$ involved?

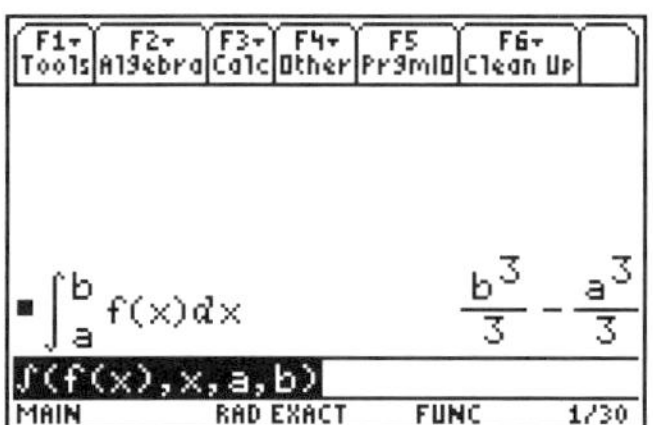

Example 2: The area under other curves

Redefine $f(x)$ and see if the relationship you observed in Example 1 works for other functions.

Solution

1. Press [2nd] [F6] **Clean Up** and select **2:NewProb** to clear variables and set other defaults.

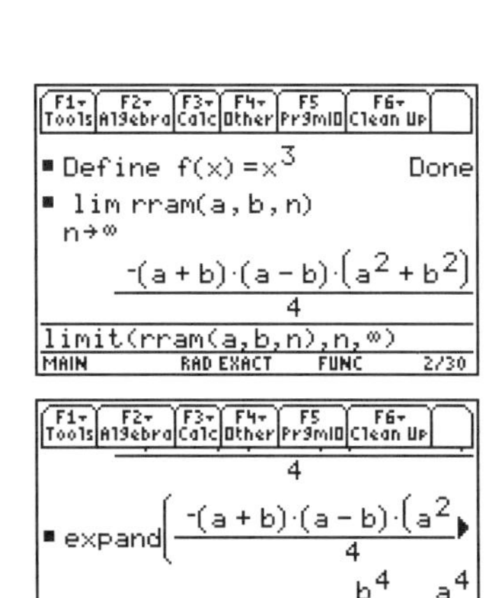

2. Use the **Define** command and define

$$f(x) = x^3$$

3. Find the limit of the right-hand Riemann Sum for the area from a to b, and compare the result with the corresponding definite integral.

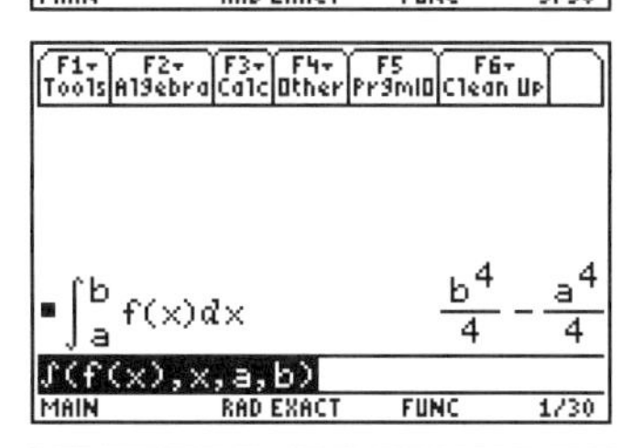

4. Define $f(x) = \cos(x)$.

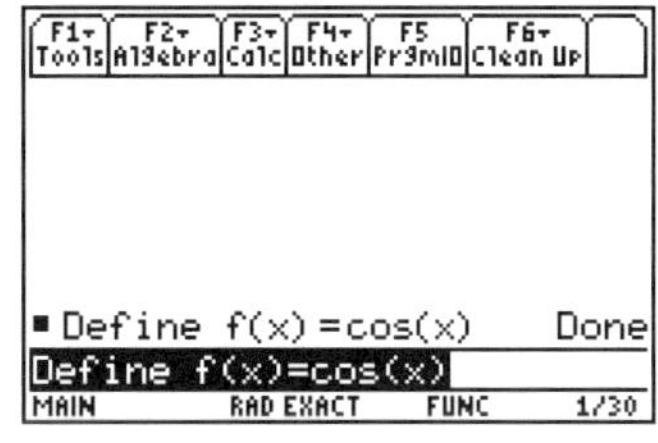

5. Find the limit of the right-hand Riemann Sum and compare the result with the corresponding definite integral.

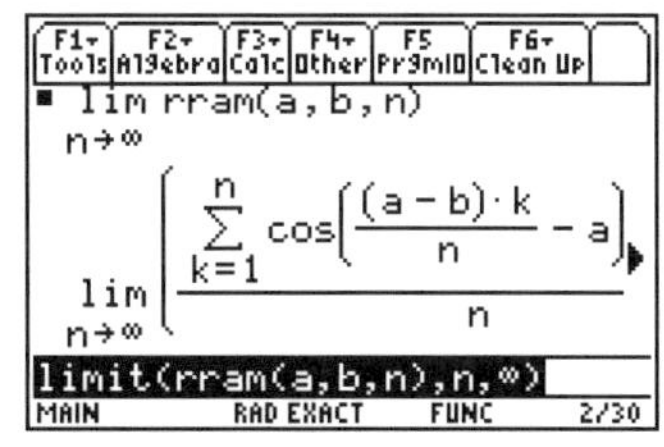

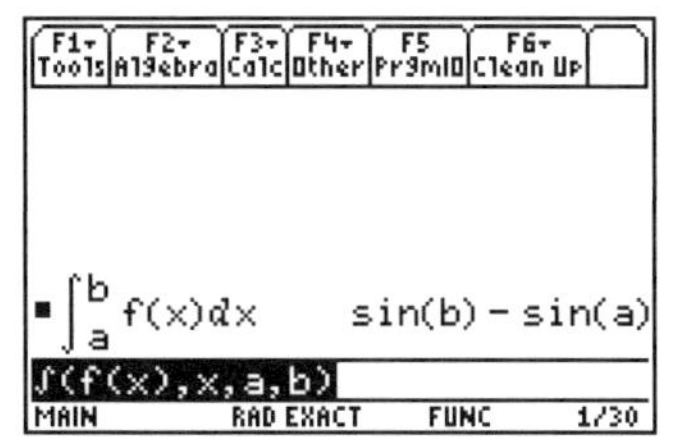

The TI-89 was not able to evaluate the limit of the Riemann Sums for $f(x) = \cos x$ since there is not a simple closed form expression for this sum; however, it was able to evaluate the definite integral. Even though the calculator could not evaluate the limit of the Riemann Sum for this function, the pattern from previous examples still holds. That is, the limit of the Riemann Sum is related to antiderivatives of $f(x)$.

Conclusion

In general, the limit of the Riemann Sums for the area bounded by a positive function $f(x)$, the vertical lines $x = a$ and $x = b$, and the x-axis is equal to $f(b) - f(a)$ where $f(x)$ is an antiderivative of $f(x)$. This result is called the Fundamental Theorem of Calculus.

Since this area is given by the definite integral

$$\int_a^b f(x)dx$$

the Fundamental Theorem also tells us that

$$\int_a^b f(x)dx = f(b) - f(a)$$

where $f(x)$ is an antiderivative of $f(x)$. This is a powerful shortcut.

Rectangular Area Approximation Program

Here is the program used to draw and evaluate the rectangular methods in this chapter. You must enter the function that bounds the area as $y1$ in the Y=Editor and enter the proper viewing Window variable values before running the program.

```
area()
Prgm
Prompt  a
Prompt  b
Input  "no. of subint=",n
(b-a)/n→h
h/2→d
Ø→l
Ø→m
Ø→r
ClrDraw
DispG
Ø→j
a→x
While  j<n
y1(x)+l→l
Line  x,Ø,x,y1(x)
Line  x,y1(x),x+h,y1(x)
Line  x+h,Ø,x+h,y1(x)
x+h→x
j+1→j
EndWhile
Pause
Disp  "left=",h*l*1.Ø
Pause
ClrDraw
DispG
Ø→j
a+h→x
While  j<n
y1(x)+r→r
Line  x-h,Ø,x-h,y1(x)
Line  x-h,y1(x),x,y1(x)
Line  x,Ø,x,y1(x)
x+h→x
j+1→j
EndWhile
Pause
Disp  "right=",h*r*1.Ø
Pause
ClrDraw
DispG
Ø→j
a+d→x
While  j<n
y1(x)+m→m
Line  x-d,Ø,x-d,y1(x)
Line  x-d,y1(x),x+d,y1(x)
Line  x+d,y1(x),x+d,Ø
x+h→x
j+1→j
EndWhile
Pause
h*m→m
Disp  "midpoint=",m*1.Ø
EndPrgm
```

Exercises

1. Evaluate **lram, rram,** and **mram** by hand for

   $$f(x) = 2x^2 + 1,\ a = 1,\ b = 2, \text{ and } n = 4.$$

2. Use the TI-89 to evaluate the rectangular approximations in Exercise 1.
3. Use the TI-89 to take the limits of the rectangular methods from Exercise 1 as $n \to \infty$.
4. Use a definite integral to evaluate the limits of the rectangular methods from Exercise 1 as $n \to \infty$. Compute by hand and then confirm with the TI-89.
5. Evaluate **lram, rram,** and **mram** by hand for

   $$f(x) = \cos(x),\ a = 0,\ b = \pi, \text{ and } n = 4.$$

6. Use the TI-89 to evaluate the rectangular approximations in Exercise 5.
7. Use the TI-89 to take the limits of the rectangular methods from Exercise 5 as $n \to \infty$.

   Use a definite integral to evaluate the limits of the rectangular methods from Exercise 5 as $n \to \infty$. Compute by hand and then confirm with the TI-89.

Chapter 6

Applications of Integrals

In this chapter, you will use the TI-89 to investigate various applications of integration. In some cases, you will use a symbolic approach; but in other cases where there may be no closed analytic solution, you will use graphical and numerical methods.

Example 1: Area between two curves

Find the area bounded by $y = e^{x^2} - 2$ and $y = \cos(x)$.

Solution

First, graph the two equations. The area is given by

$$\int_a^b y2(x) - y1(x)dx$$

Then use the graph to get the values for the left and right intersection points. With these values, you can evaluate the definite integral.

1. Press [2nd] [F6] **Clean Up** and select **2:NewProb** to clear variables and set other defaults.
2. Set **Exact/Approx=AUTO** on Page 2 of the MODE dialog box.
3. In the Y= Editor, enter the equations in $y1$ and $y2$.
4. Set the Window variable values as shown for a [-2,2] x [-2,2] viewing window.
5. Graph the equations.

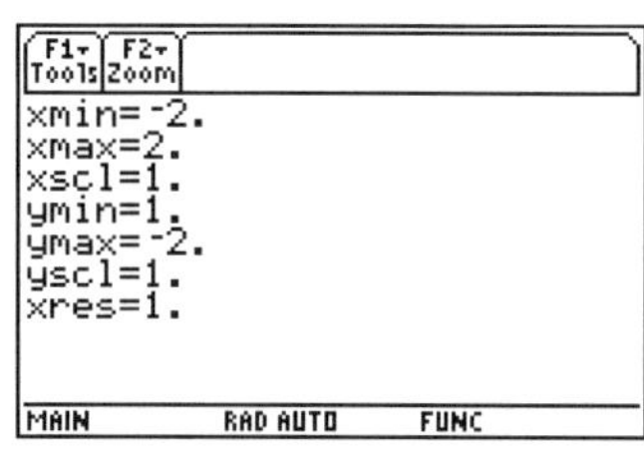

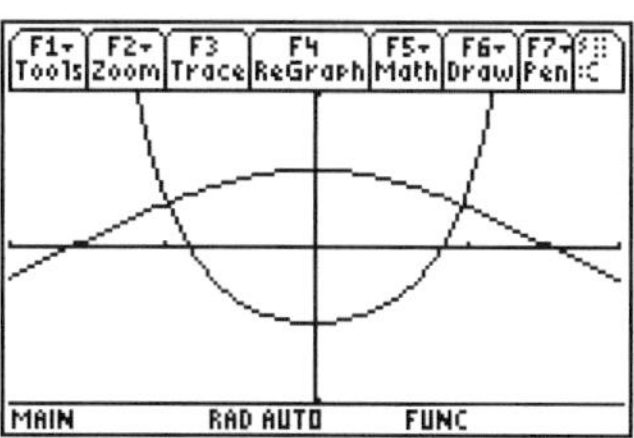

6. Find the left point of interaction. Press [F5] **Math** and select **5:Intersection**. Press [ENTER] to use $y1$ as the first curve, and press [ENTER] again to use $y2$ as the second curve. Press ◀ and ▶, or type values, to set the bounds.

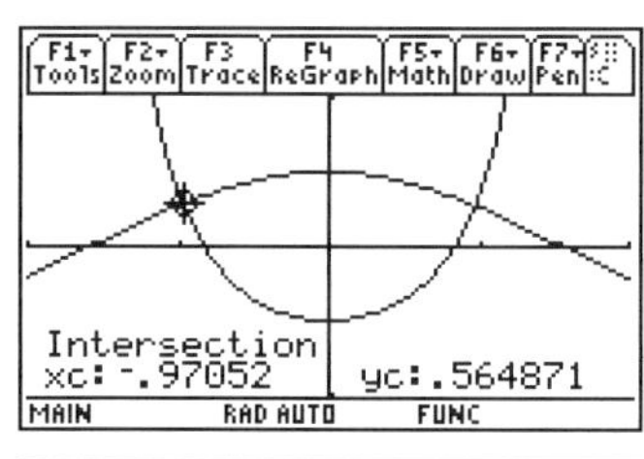

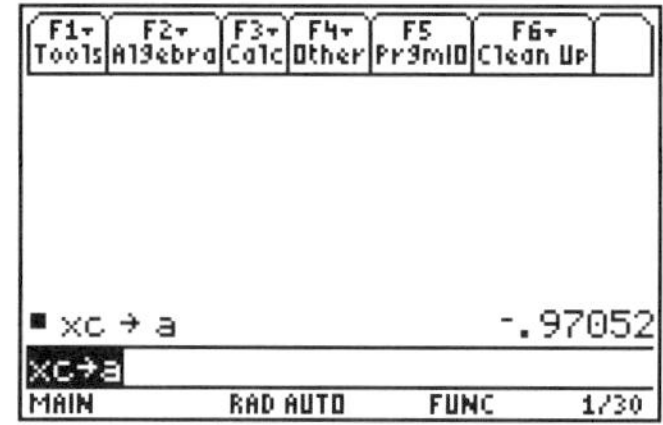

7. Press [HOME] to return to the Home screen and store the x-coordinate as a for later use in the definite integral.

8. Return to the graph screen and repeat steps 6 and 7 to find and store the right point of intersection.

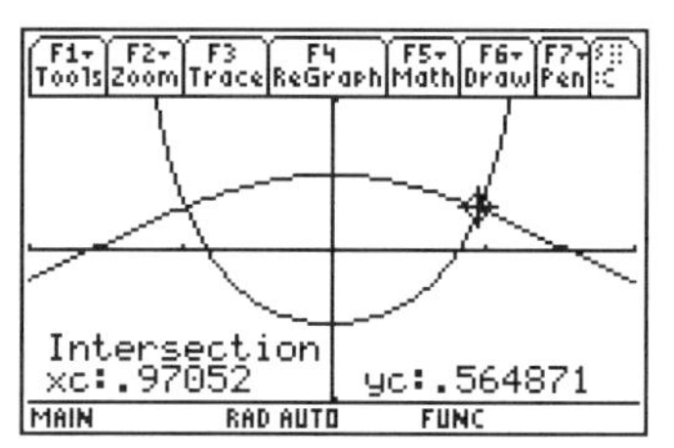

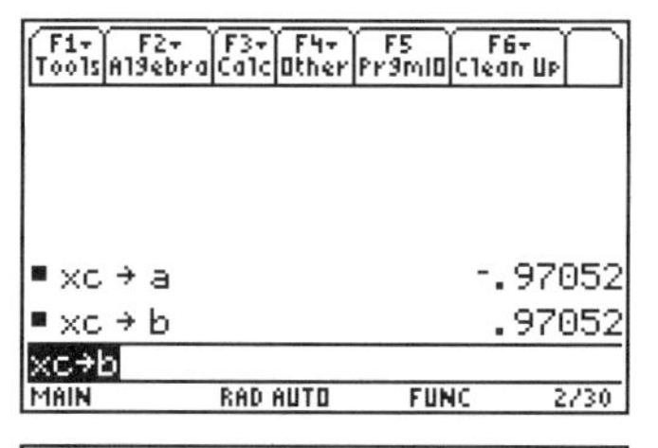

9. Evaluate the definite integral.

[2nd] [∫] **Y2** [(] **X** [)] [−] **Y1** [(] **X** [)] [,] **X** [,] **A** [,] **B** [)] [ENTER]

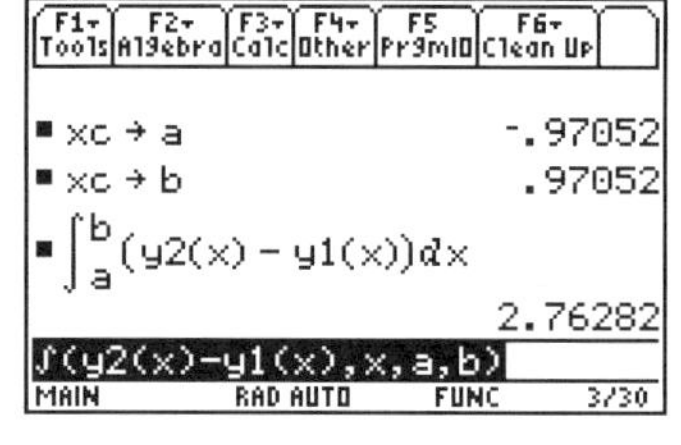

The area between the two curves is about 2.76282.

Example 2: Arc length

Find the length of the astroid $y = (1 - x^{\frac{2}{3}})^{\frac{3}{2}}$ in the first quadrant.

Solution

First, find the length of a curve using the built-in Arc feature of the TI-89. Then compare the result with the definite integral for arc length, where arc length is

$$\int_a^b \sqrt{1 + (\frac{dy}{dx})^2}\, dx$$

1. Press [2nd] [F6] **Clean Up** and select **2:NewProb** to clear variables and set other defaults.

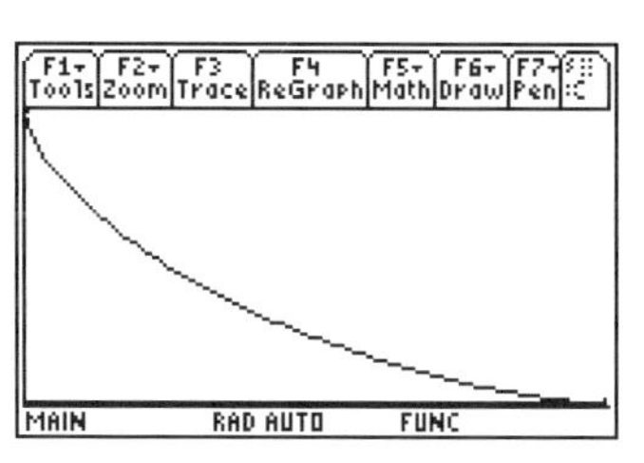

2. Enter the equation in $y1$ in the Y= Editor and clear $y2$. Set the Window variable values for a [0,1] x [0,1] viewing window. Then graph $y1$.

3. Press [F5] **Math** and select **8:Arc**. You are prompted for the x-coordinates of the left and right endpoints of the arc.

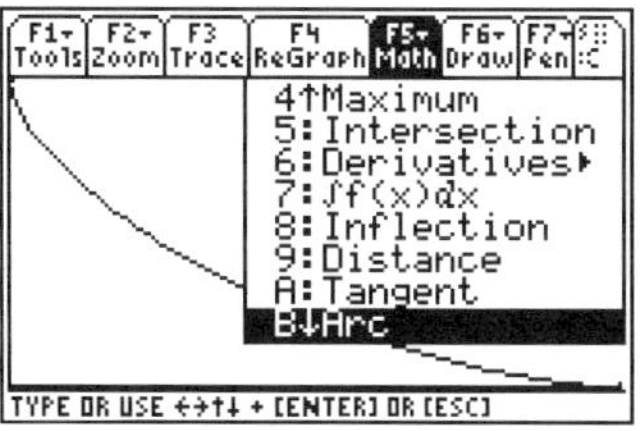

4. Enter **0** for the x-coordinate of the first point *(xc)* and **1** for the x-coordinate of the second point. You can ignore the y-coordinate values *(yc)*.

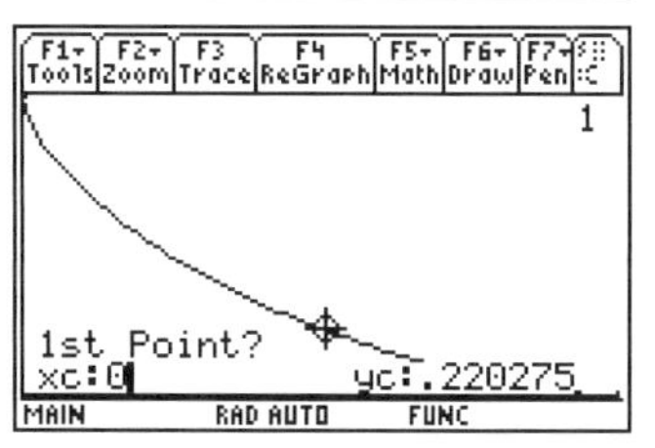

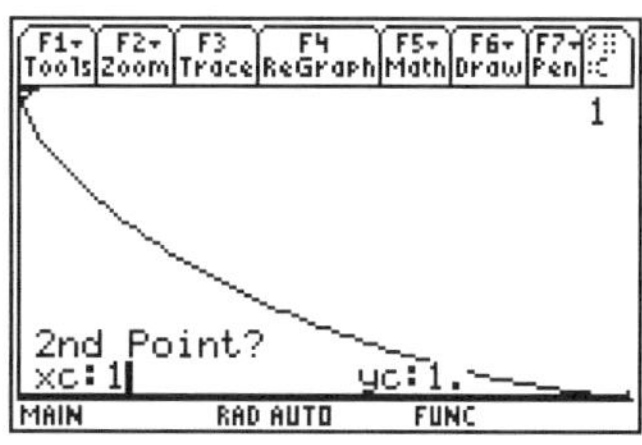

The length of the arc is 1.5. Now compare this result with the definite integral result.

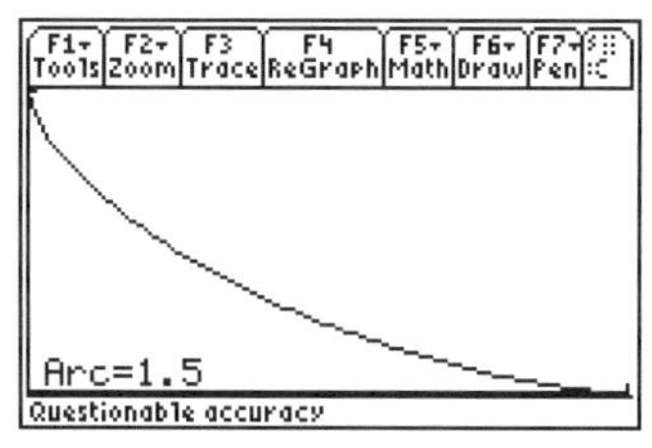

5. Store the derivative of *y1* in the variable $dydx$ to make it easier to enter the definite integral.

   [2nd] [d] **Y1** [(] **X** [)] [,] **X** [)] [STO▶] **DYDX** [ENTER]

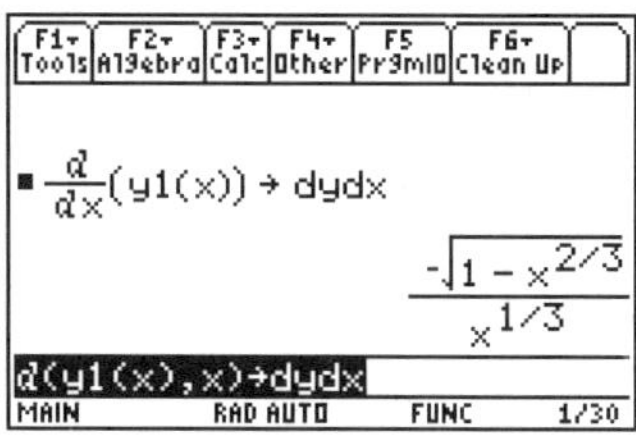

6. Enter the expression for the definite integral.

   [2nd] [∫] [2nd] [√] **1** [÷] **DYDX** [^] **2** [,] **X** [,] **0** [,] **1** [)] [ENTER]

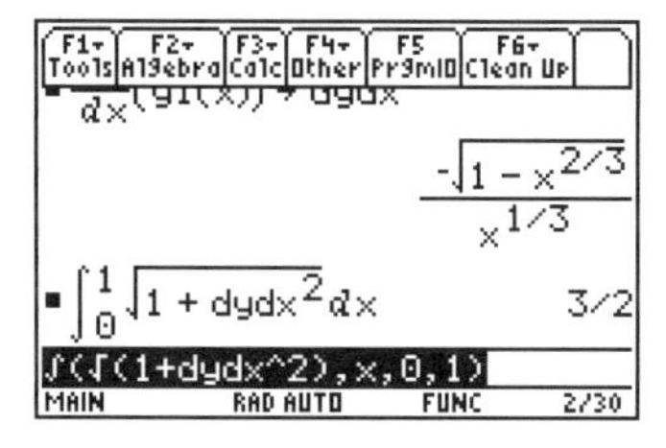

This result confirms the value we found using the **Arc** feature of the **Math** menu.

Example 3: First-order differential equations

The next few simple examples illustrate applications of integration that can be solved using the Differential Equation Solver **(deSolve)** of the TI-89.

Solve the differential equation

$$y' = 9.8t + 2.7$$

Solution

1. Press [2nd] [F6] **Clean Up** and select **2:NewProb** to clear variables and set other defaults.

2. Find the solution using the **deSolve(** command.

   [CATALOG] **deSolve(Y** [2nd] ['] [=] **9.8 T** [+] **2.7** [,] **T** [,] **Y** [)] [ENTER]

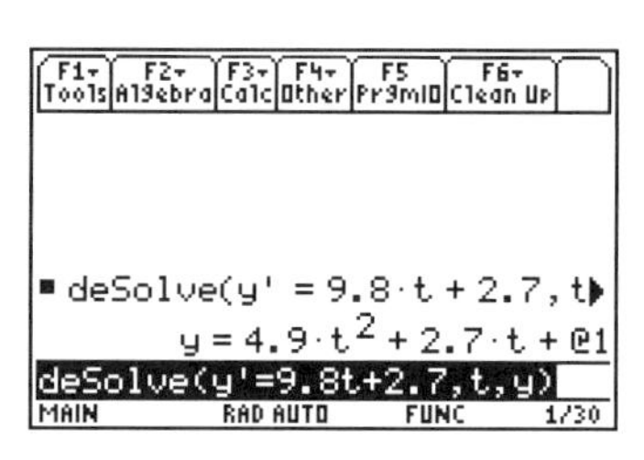

   The solution is

$$y = 4.9t^2 + 2.7t + C$$

   Notice that the constant of integration is represented with the @ 1 symbol.

3. To solve the same differential equation with initial conditions $y(0) = 7$:

   Edit the **deSolve(** command as follows:

   deSolve(y′=9.8t+2.7 and y(0)=7,t,y)

   Note: The **and** operator is in the CATALOG.

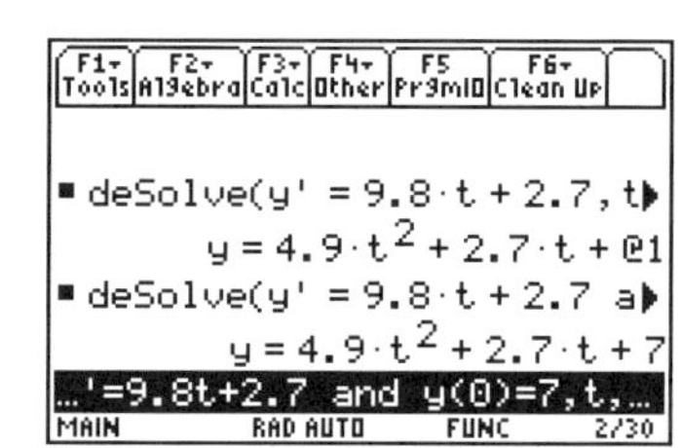

Example 4: Second-order differential equations

A ball was tossed straight up from an initial height of 0.29 meters and with an initial velocity of 3.8 m/s. Solve the second-order differential equation

$$y'' = 9.8, \quad y(0) = .29, \quad y'(0) = 3.8$$

to find an equation to model the height of the ball over time.

Solution

Clear the calculator and enter the **deSolve(** command.

[CATALOG] **deSolve(Y** [2nd] ['] [2nd] ['] [=] [(-)] **9.8** [CATALOG] **and Y** [(] **0** [)] [=] **0.29** [CATALOG] **and Y** [2nd] ['] [(] **0** [)] [=] **3.8** [,] **T** [,] **Y** [)]

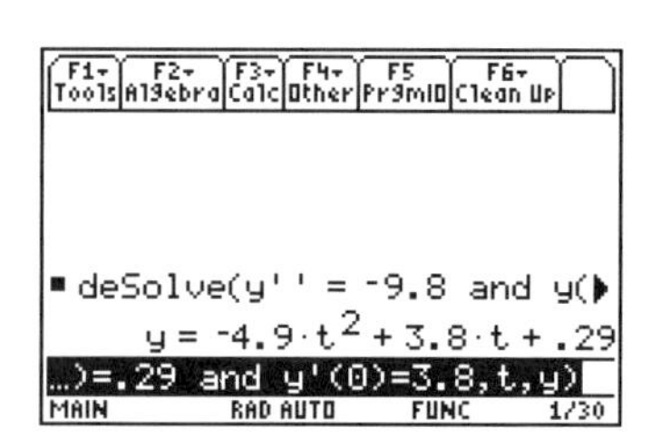

Example 5: Scatter plots and regression curves

We measured the height of the ball in Example 4 with a Texas Instruments CBL™. Here is the data from the CBL.

Time (seconds)	Height (meters)
0	.29
.04	.44
.08	.57
.12	.68
.16	.78
.20	.86
.24	.92
.28	.96
.32	.99
.36	1.0

Continued in next column

Time (seconds)	Height (meters)
.40	1.0
.44	1.0
.48	.97
.52	.94
.56	.89
.60	.82
.64	.74
.68	.64
.72	.53
.76	.40

Make a scatter plot of the data and compare the solution found in Example 4 with the data.

Solution

1. Press [2nd] [F6] **Clean Up** and select **2:NewProb** to clear variables and set other defaults.

2. Start the Data/Matrix Editor by pressing [APPS] and selecting **6:Data/Matrix Editor**. Then select **3:New.**

3. Move the cursor down to **Variable** and enter a name for the data variable (**ball**).

4. Press [ENTER] twice to display the Data/Matrix Editor. Enter the data for time in column 1 (*c1*) and the data for height in column 2 (*c2*). The headings above *c1* and *c2* are optional.

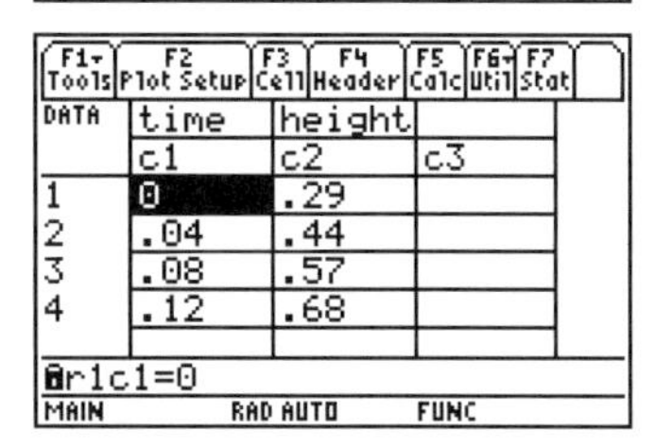

5. Set up the scatter plot. Press [F2] **Plot Setup** to display the main\ball dialog box. Press [F1] to define the plot. For the **Plot Type,** press ▶ and select **1:Scatter**. For **Mark**, select **1:Box.** Enter **c1** and **c2** for x and y.

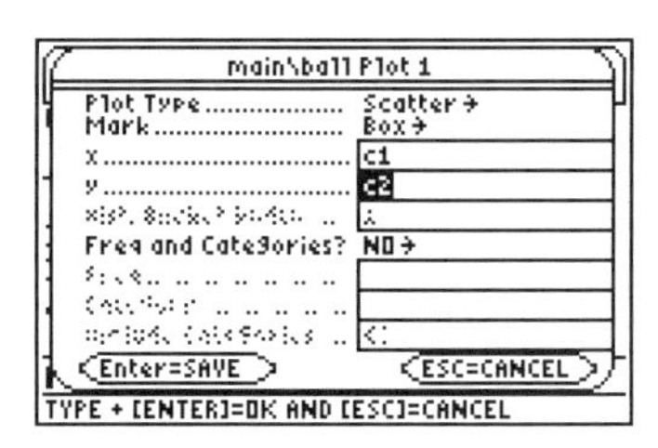

6. Press [ENTER] twice to return to the Data/Matrix Editor.

7. Press ◆ [Y=] to display the Y = Editor. In $y1$, enter the equation found with **deSolve** in Example 4. Use x in place of t.

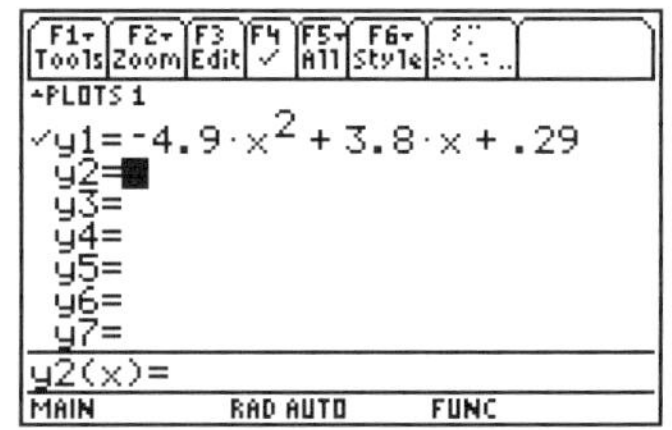

8. Press [F2] **Zoom** and select **9:ZoomData** to see the match between the equation and the data.

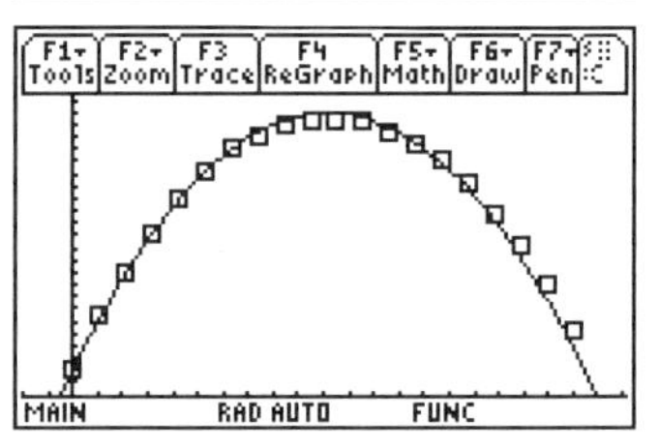

9. Finally, find a quadratic regression equation for the data and compare it with the equation found with **deSolve(**. Press [APPS] **6:Data/Matrix Editor** and then select **1:Current** to return to the Data/Matrix Editor.

10. Press [F5] **Calc** to display the main\ball Calculate dialog box. Press ▶ and select **9:QuadReg**.

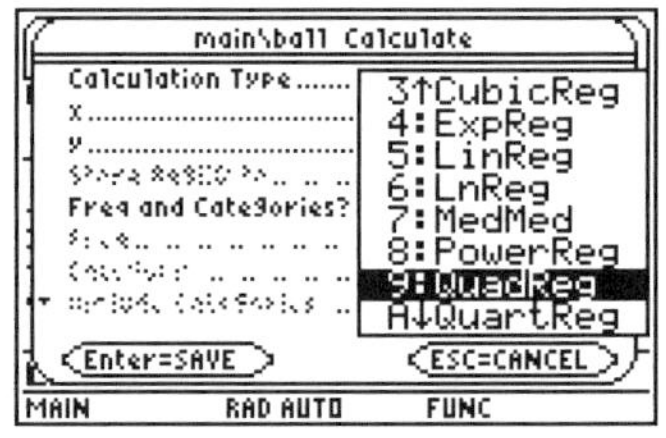

11. Enter **c1** and **c2** for x and y. Then move to the **Store RegEQ to** (store regression equation) menu, press ▶ and select **y2(x)**.

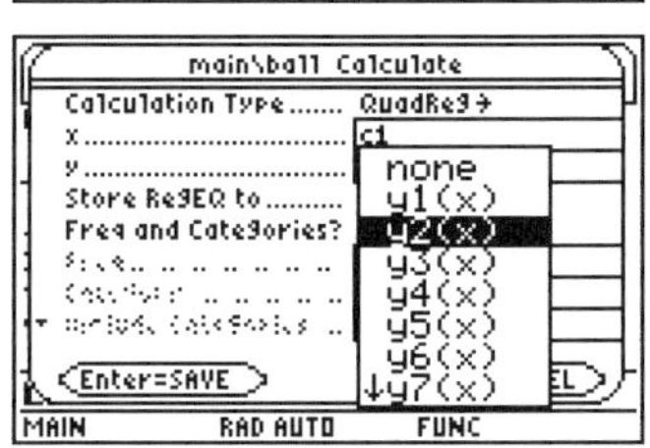

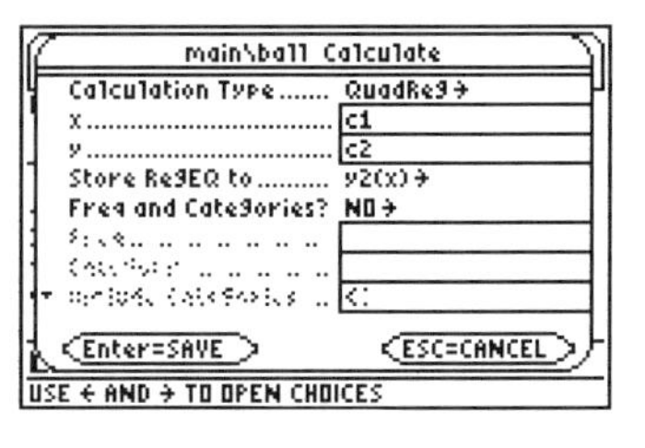

12. Press [ENTER] to calculate and store the regression equation. Then press [ENTER] [◆] [GRAPH] to see and compare the graph of the regression equation with the graph of the equation found with **deSolve(**.

    The two equations match fairly well.

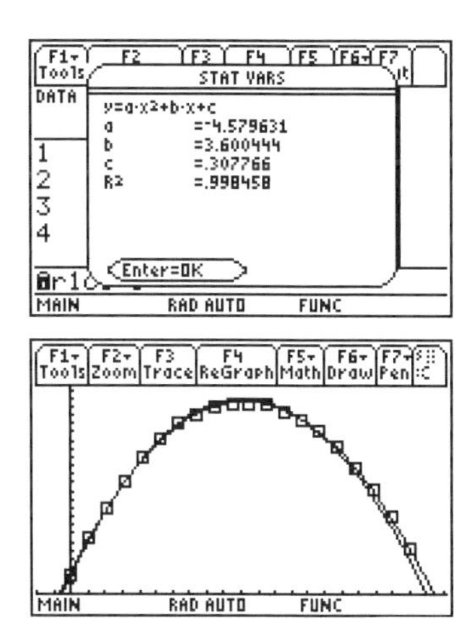

Example 6: Newton's Law of Cooling

This example uses **deSolve(** to solve a problem involving Newton's Law of Cooling. Newton's law describes the rate at which an object cools when it is immersed in surroundings that are colder than the object. It says the rate at which the object's temperature changes is directly proportional to the difference between the temperature of the object and the temperature of the surrounding medium.

If y is the temperature of an object over time t and t_s is the temperature of the surroundings, then Newton's law says

$$\frac{dy}{dt} = -k(y - t_s).$$

A temperature probe is connected to a Texas Instruments CBL™. The probe is heated to a temperature of 65°C. It is placed in water that has a temperature of 5°C. The probe cools to a temperature of 11°C in 30 seconds. Predict the temperature 60 seconds after the probe was placed in the cold water.

Solution

1. Press [2nd] [F6] **Clean Up** and select **2:NewProb** to clear variables and set other defaults.

2. Enter the **deSolve(** command.

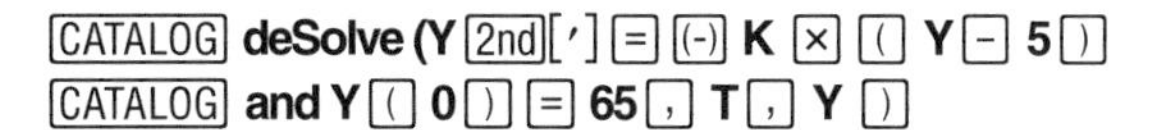

[CATALOG] **deSolve (Y** [2nd][′] [=] [(-)] **K** [×] [(] **Y**[−] **5**[)]
[CATALOG] **and Y**[(] **0**[)] [=] **65**[,] **T**[,] **Y** [)]

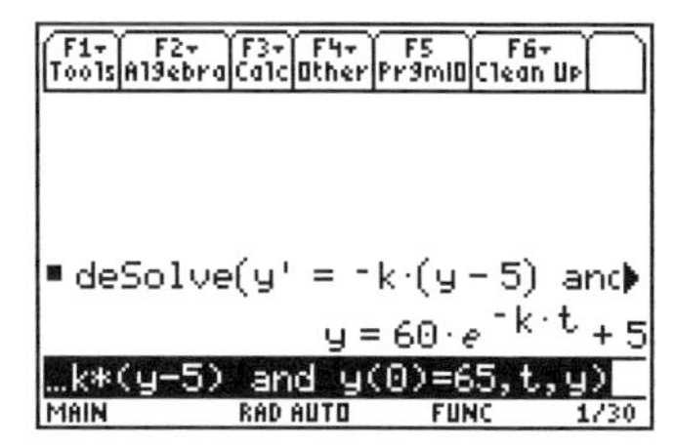

3. Now use the result and final conditions in the **solve(** command to find the constant of proportionality k.

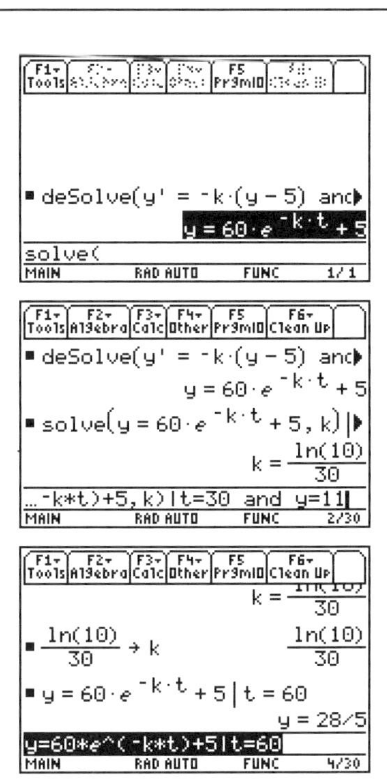

4. Store this value in k.

   [2nd] [LN] **10** [)] [÷] **30** [STO▸] **K**

5. Evaluate the equation from the result of step 2 for temperature of t=60.

   ⊖ ⊖ ⊖ ⊖ ⊖ [ENTER] [|] **T** [=] **60** [ENTER]

The temperature should be about 28/5 or 5.6°C after 60 seconds.

Example 7: Resistance proportional to velocity

A ball is dropped from a height of 75 meters. Assume the acceleration due to gravity is -9.8 m/s^2 and deceleration due to air resistance is directly proportional to velocity with the constant of proportionality equal to 0.05. When will the ball hit the ground?

Solution

The distance traveled can be found by solving the differential equation

$$y'' = {}^{-}9.8 - .05y' \text{ with initial conditions}$$

$$y(0) = 75 \text{ and } y'(0) = 0.$$

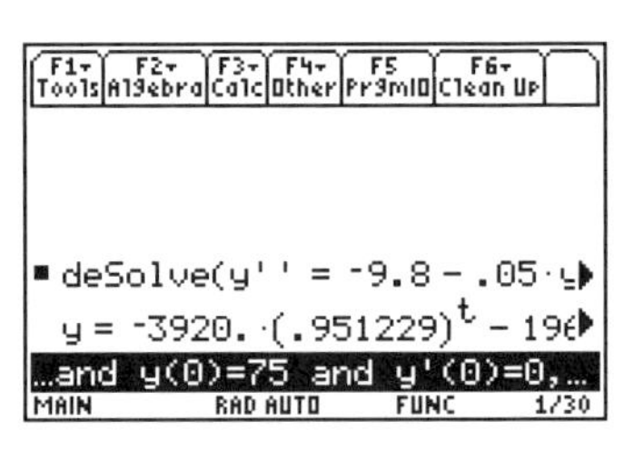

1. Press [2nd] [F6] **Clean Up** and select **2:NewProb** to clear variables and set other defaults.

2. Enter the **deSolve(** command using the equation and initial conditions.

   [CATALOG] **deSolve(y** [2nd] [′] [2nd] [′] [=] [(-)] **9.8** [−] **.05Y** [2nd] [′] [CATALOG] **and Y** [(] **0** [)] [=] **75** [CATALOG] **and Y** [2nd] [′] [(] **0** [)] [=] **0** [,] **T** [,] **Y** [)]

The expression for distance is

$y = -3920(.951229)^t + 196t + 3995$.

3. Use this result with the **solve(** command to solve for t with $y = 0$.

   The ball will hit after about 4.04412 seconds.

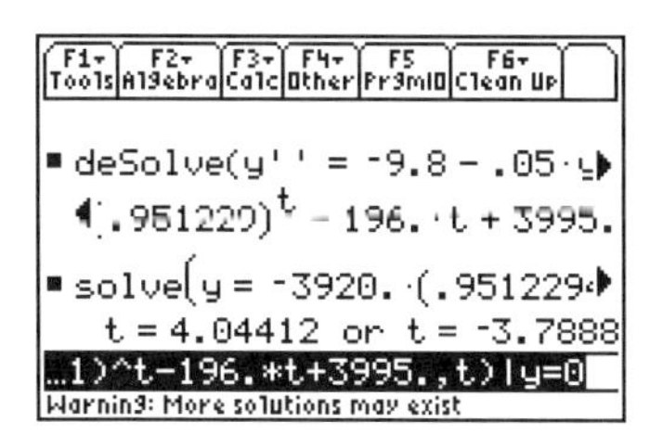

Example 8: Logistic growth

In the last example, you will solve a logistic growth problem. In logistic growth problems, assume that the rate of growth of a population (k) is directly proportional to both the population (y) and the carrying capacity (C) minus the population.

$$\frac{dy}{dt} = ky(C - y)$$

Eight wolves are introduced into a national park. Assume zoologists have determined a carrying capacity of 250 wolves and a growth rate constant of .001. When will the population reach 100?

Solution

1. Press [2nd] [F6] **Clean Up** and select **2:NewProb** to clear variables and set other defaults.

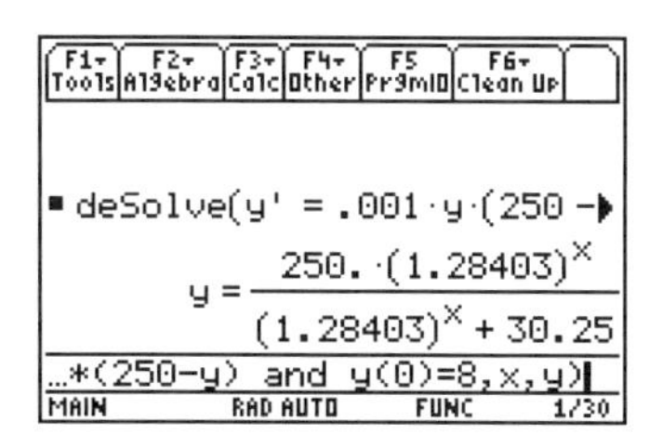

2. Use **deSolve(** to solve the differential equation. Let x equal time so that you can graph the solution.

   [CATALOG] **deSolve(Y** [2nd] ['] [=] **.001Y** [×] [(] **250** [-] **Y** [)] [CATALOG] **and Y** [(] **0** [)] [=] **8** [,] **X** [,] **Y** [)]

3. Define $y1(x)$ to be the solution from step 2.

   [CATALOG] **Define Y1** [(] **X** [)] [=] ⊙ [ENTER]

   Delete **y=** from the expression, and press [ENTER].

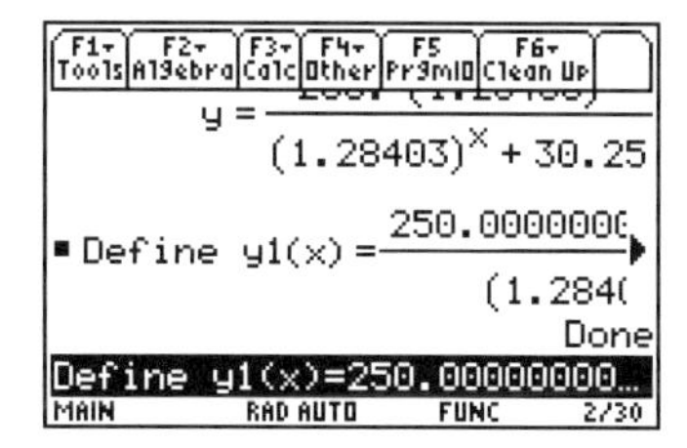

4. Set the Window variable values for a [0,50] x [0,275] viewing window.

5. Graph $y1(x)$ and trace ([F3]) until the y-coordinate is about 100.

   The population will reach 100 in about 12 years.

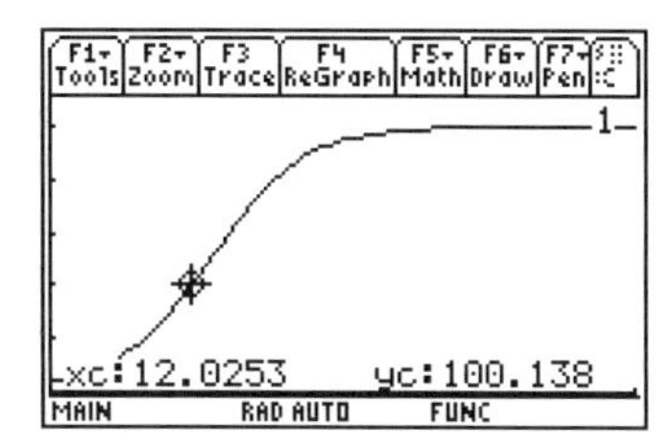

Exercises

1. Find the area between the curves

$$y = \sin(x^2) \text{ and } y = 1 - x^2.$$

2. Find the length of the curve $y = \sin^{-1} x$ for $-1 \le x \le 1$. Use the Arc feature and compare the answer with the result of the definite integral that gives arc length.

Use **deSolve(** to solve the following differential equations (exercises 3-6).

3. $y' = x^2 \cos(x)$, $y(0) = 0$

4 $v' = -\dfrac{k}{m} v$, $v(0) = v_o$

5. $y'' = -32$, $y(0) = 19$, $y'(0) = 0$

6. $q'' = 1 - 2q - 2q'$, $q(0) = 0$, $q'(0) = 0$

7. A book was dropped from a height of .8649 meters. Its height above ground was measured with a Texas Instruments CBL™. Here are the data:

Time (seconds)	Height (meters)
0	.8649
.02	.8583
.04	.8484
.06	.8364
.08	.8188
.10	.7991
.12	.7749
.14	.7475
.16	.7156
.18	.6805
.20	.6421
.22	.6004
.24	.5543
.26	.5060
.28	.4544

(a) Solve the second-order differential equation

$$y'' = -9.8$$

to find an equation to model the height of this book over time.

(b) Make a scatter plot of the data and graph the solution from part (a) with this scatter plot.

8. A cup of hot chocolate is left to cool on a kitchen table. It cools from 93°C to 55°C in 15 minutes. If the room temperature is 21°C, predict the temperature of the hot chocolate 20 minutes after it is placed on the table.

9. A projectile is fired straight up with an initial velocity of 87m/s. Assume the acceleration due to gravity is -9.8 m/s^2 and deceleration due to air resistance is directly proportional to velocity with the constant of proportionality equal to 0.05. When will the projectile hit the ground?

10. A rumor spreads through a school with 1500 students. If 4 students initially hear the rumor and the growth rate is .00025, use a logistic model to predict how long it will take for the rumor to spread to 1000 students.

Chapter 7

Differential Equations and Slope Fields

In Chapter 6, you used the **deSolve(** command to solve differential equations analytically. In this chapter, you will use the differential equation graphing mode of the TI-89 to solve differential equations graphically and numerically. You will start with a solution that should look familiar and then move to more difficult equations.

Before you begin this chapter, reset your TI-89 to the default settings by pressing [2nd] [MEM] [F1] **Reset,** then select **3:Default** [ENTER] [ENTER].

Example 1: Slope fields and specific solutions

Solve this differential equation.

$$y' = \frac{x}{4}$$

Solution

The analytic solution is

$$y = \frac{x^2}{8} + C$$

As you choose various values for C, you will obtain parabolas with the same shape that are shifted vertically from each other. You can graph this family of parabolas with the differential equation graphing mode on the TI-89.

1. Press [MODE] to display the MODE dialog box. Press ▶ and select **6:DIFF EQUATIONS**. Press [ENTER] to save the change and return to the Home screen.

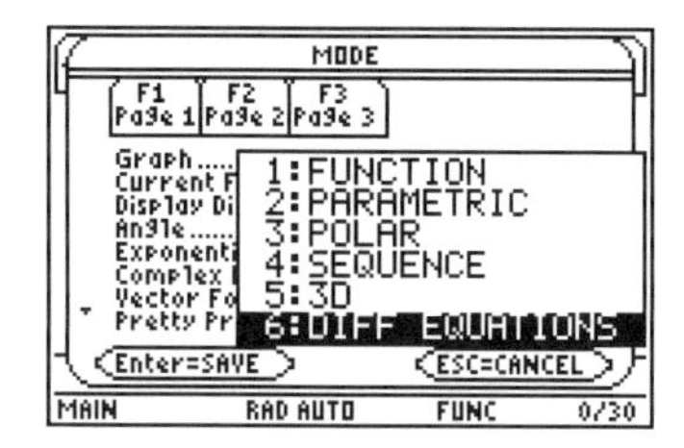

2. Press [♦] [Y=] to display the Y= Editor. Clear any equations that are stored.

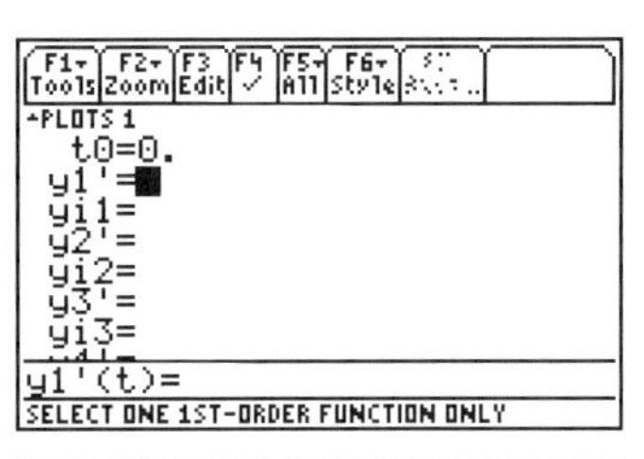

3. Enter the differential equation using t in place of x.

   T [÷] **4** [ENTER]

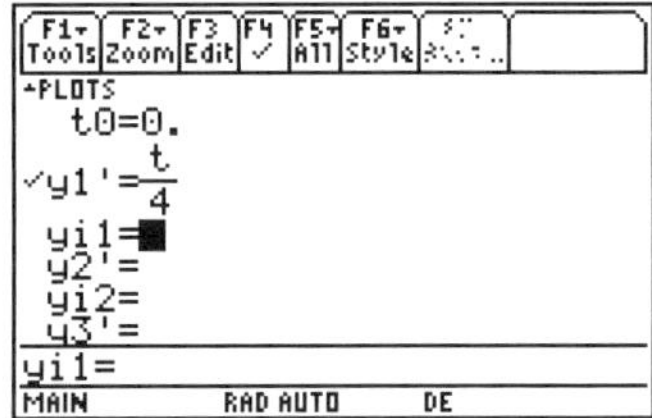

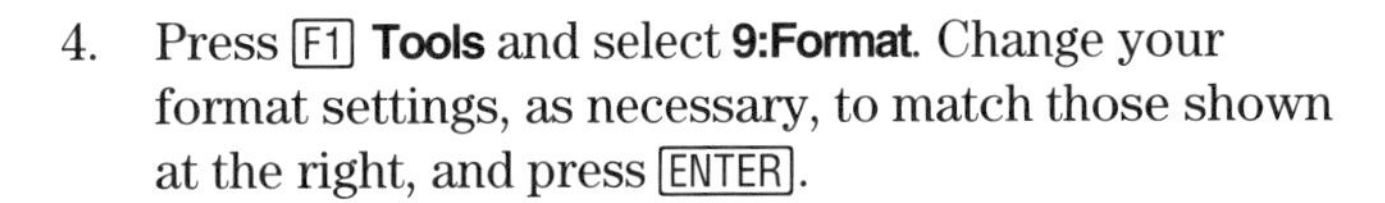

4. Press [F1] **Tools** and select **9:Format**. Change your format settings, as necessary, to match those shown at the right, and press [ENTER].

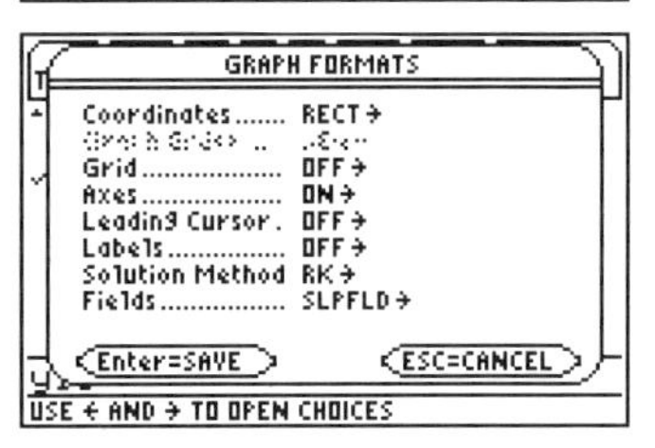

5. Graph the solution in a Zoom Decimal window by pressing [F2] **Zoom** and selecting **4:ZoomDec**. This graph is called a *slope field*. It suggests to the imagination a family of the parabolic solutions to the differential equation.

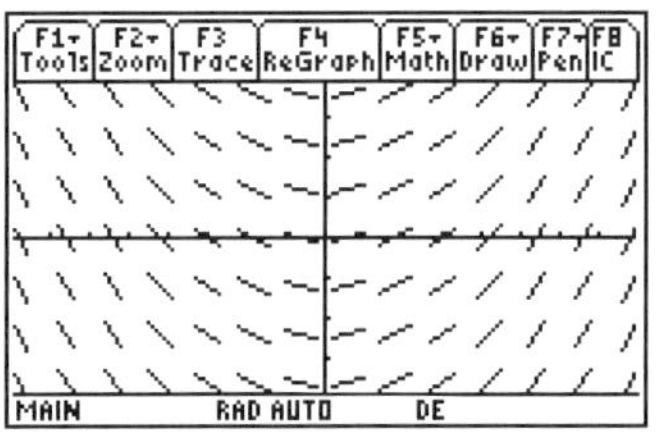

6. You can see specific members of this family of solutions by specifying initial conditions for the differential equation. If you want to see the specific graph corresponding to the initial condition $y(0) = 0$, return to the Y= Editor and enter **0** for $yi1$.

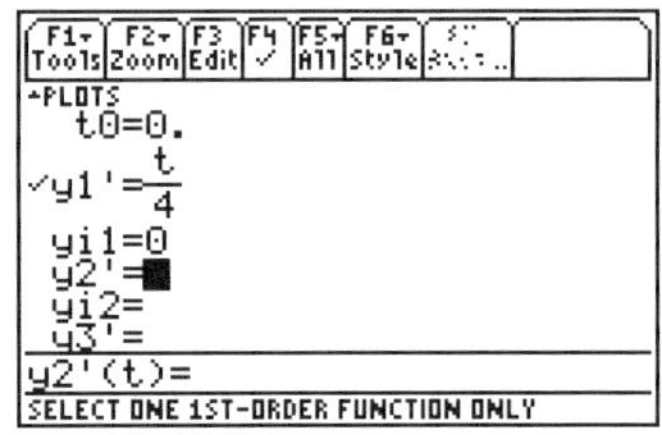

7. Press [♦] [GRAPH] to see the specific solution graphed with the slope field.

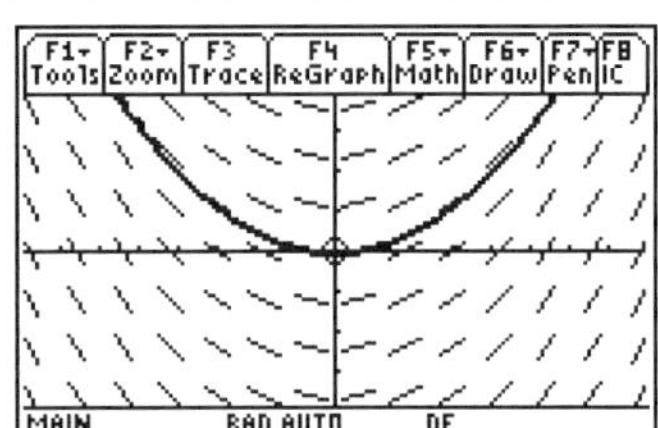

8. To see several members of the family of solutions corresponding to $y(0) = -1$, $y(0) = 0$, and $y(0) = 1$, enter a list of initial conditions in the Y= Editor.

   [2nd] [{] [(-)] **1** [,] **0** [,] **1** [2nd] [}] [ENTER]

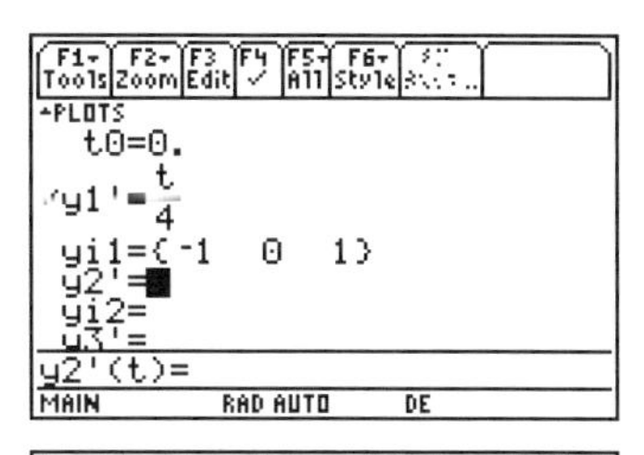

9. Press [◆] [GRAPH]. This graph gives more insight into the slope field. If you start at any initial point in the slope field and move so that the line segments in the field are tangent to your path, you will trace out a specific solution to the differential equation.

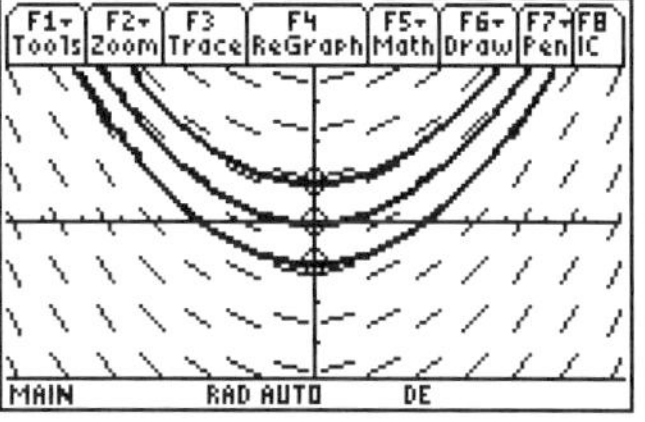

Example 2: Window variables and initial conditions

Graph the solution to

$$y' = e^{-x^2}, \quad y(-3) = 0$$

Solution

1. In the Y= Editor, enter the differential equation as $y1'$ using t in place of x. Then enter the initial value of x in *t0* and the initial value of y in $yi1$.

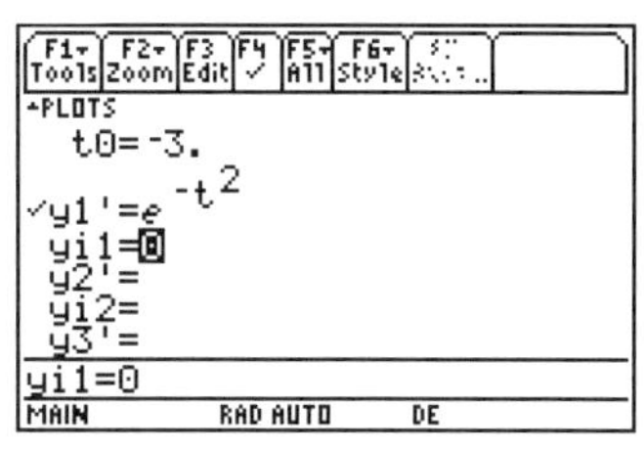

2. Select a [-3,3] x [0,2] viewing window. Display the Window Editor by pressing [◆] [WINDOW]. Enter the values for the Window variables as shown.

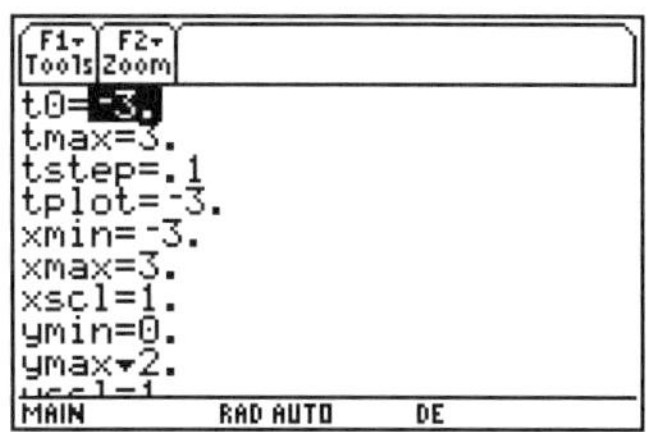

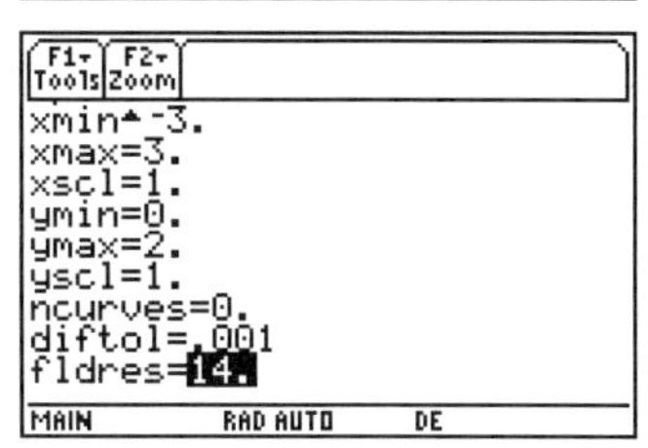

The Window variables shown in these examples have the following functions:

Variable	Function
tmin	The initial *x*-coordinate. This is the point at which the TI-89 will begin calculating the solution.
tmax	The point at which the TI-89 will stop calculating and graphing the solution (when the slope field is turned off)
tstep	The horizontal spacing between the points calculated in the solution
tplot	The point at which plotting begins (when slope field is turned off)
xmin	The left boundary of the viewing window
xmax	The right boundary of the viewing window
xscl	The spacing between the marks on the x-axis
ymin	The bottom boundary of the viewing window
ymax	The top boundary of the viewing window
yscl	The spacing between the points on the y-axis
ncurves	The number of specific curves drawn if no initial condition is specified
diftol	Affects the speed and accuracy of the solution when the format solution setting of **RK** is selected. (When Euler is selected, **diftol** is replaced with **Estep**.)
fldres	Affects the size of the segments shown in the slope field

3. Press [◆] [GRAPH] to see the solution.

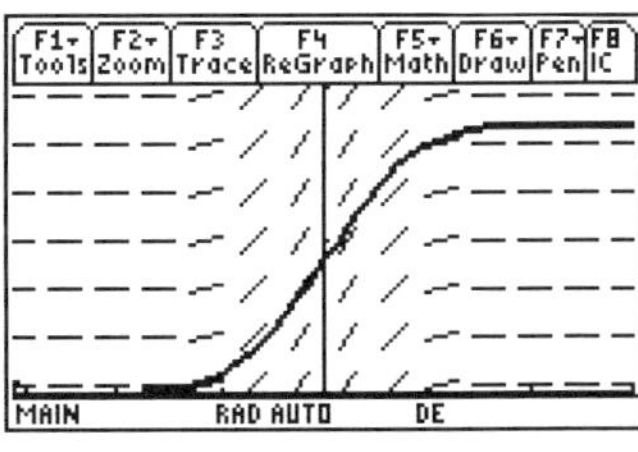

4. Turn the slope field off by pressing [F1] **Tools** and selecting **9:Format.** Set **Fields** = **3:FLDOFF**.

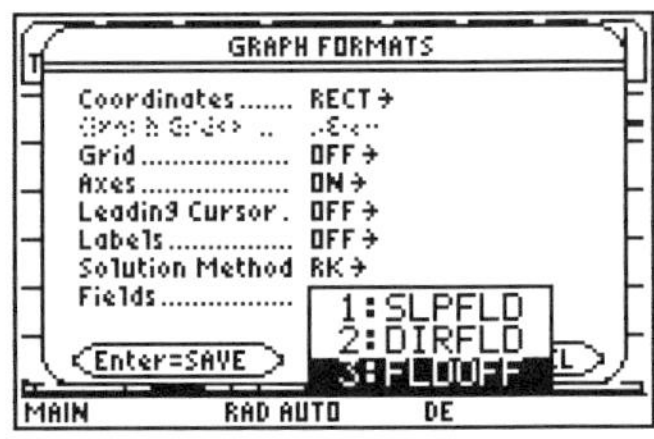

5. Press [ENTER] to graph the solution. This graph is a scaled version of a normal probability density function. Probability density functions are used in probability and statistics.

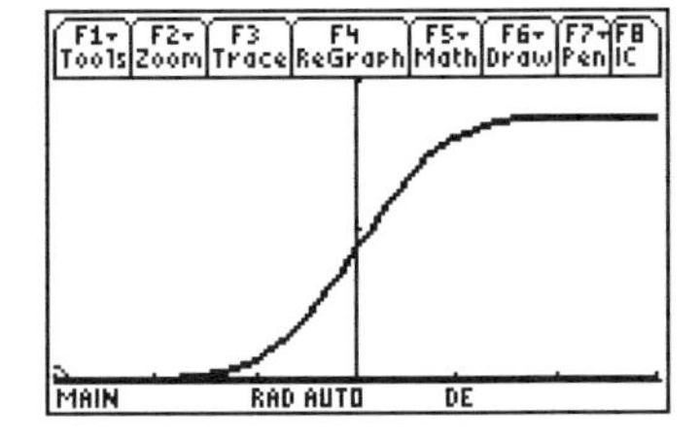

Example 3: Tabular solutions to differential equations

The differential equation that describes the decay of carbon-14 is

$$\frac{dy}{dt} = -\frac{\ln 2}{5700}y, \quad y(0) = 1$$

where y is the proportion of radioactive nuclei present in a dead organism to radioactive nuclei present in the organism while it lived and t is elapsed time in years since the organism died.

(a) Find the proportion of radioactive carbon-14 in a sample after 1000, 2000, 3000, 4000, and 5000 years.

(b) How old is a sample in which 20% of the carbon-14 nuclei have decayed?

Solution

1. Make sure **Graph** mode is set to **DIFF EQUATIONS.**

2. In the Y= Editor, enter the differential equation along with the initial conditions.

   y1′=⁻ln(2)/5700*y1
   t0=0
   yi1=1

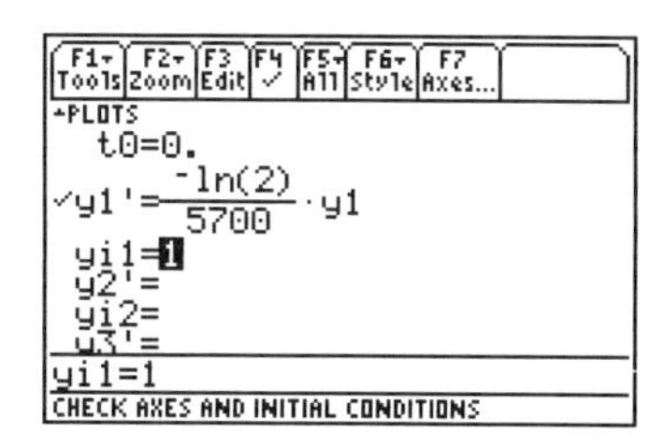

3. In the Window Editor, set **tmax= 5000**, **tstep=10**, and **xmin=0**. All other values are the same as in Example 2.

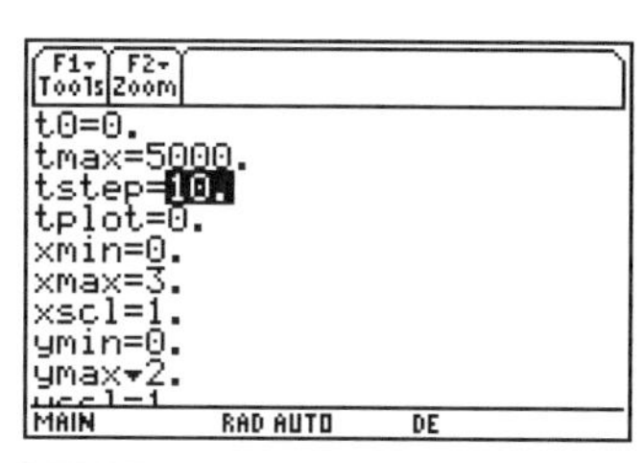

4. Press [◆] [TblSet] and set the table parameters:

   tblstart=0
   Δtbl=1000
   Graph ↔ Table = OFF
   Independent = AUTO

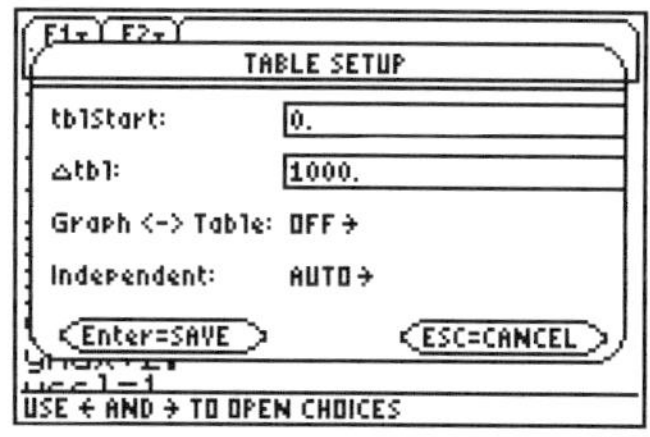

5. When you have these settings, press [ENTER] twice to save them. Then press [◆] [TABLE] to see the table of values for the solution to the differential equation. Scroll down to see the value for t=5000. The $y1$ column shows the proportion of radioactive carbon-14 after 1000, 2000, 3000, 4000, and 5000 years.

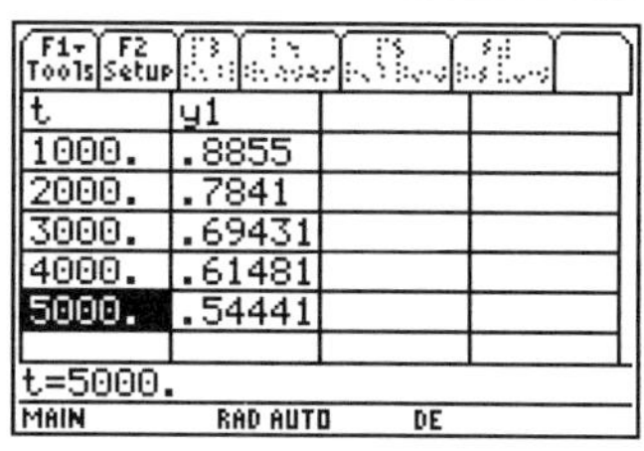

t	y1
1000.	.8855
2000.	.7841
3000.	.69431
4000.	.61481
5000.	.54441

6. You can see from the table that the answer to part (b) is somewhere between 1000 and 2000 years. Press [◆] [TblSet] to display the TABLE SETUP dialog box and start the table at 1000 years (**tblstart**) with 100-year increments (**Δtbl**).

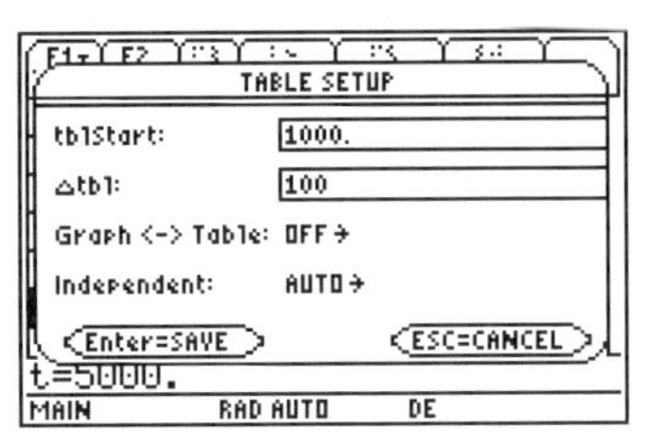

7. Press [ENTER] twice to save the changes and see the new table. Scroll down the table until the second column is close to 0.80.

   The answer is somewhere between 1800 and 1900 years.

t	y1
1500.	.83326
1600.	.82319
1700.	.81324
1800.	.80341
1900.	.79369

t=1900.

MAIN RAD AUTO DE

8. Return to the TABLE SETUP dialog box and start at 1800 with 10-year increments. This method of expanding the table in the area of interest is called *tabular zoom*.

   It appears the sample is about 1830 years old.

   The tables in this example were calculated using a solution method known as the *Runge-Kutta* (RK) method. You can select either the Runge-Kutta method or Euler's solution method in the for GRAPH FORMATS dialog box. The Runge-Kutta method is generally more accurate but, Euler's method is often faster. Consult a calculus book to see the details involved in these methods.

t	y1
1800.	.80341
1810.	.80243
1820.	.80145
1830.	.80048
1840.	.79951

t=1800.

MAIN RAD AUTO DE

9. Change to Euler's solution method by pressing [◆] [Y=] [F1] **Tools** and selecting **9:Format**. Set **Solution Method = EULER**.

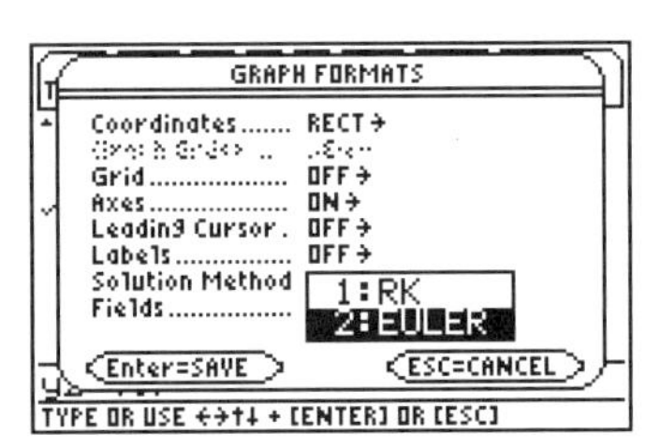

10. Press [ENTER] to save the **Euler** setting. Then press [◆] [TABLE] to see the table calculated with Euler's method. Notice the differences between this table and the table generated with the RK method in step 8.

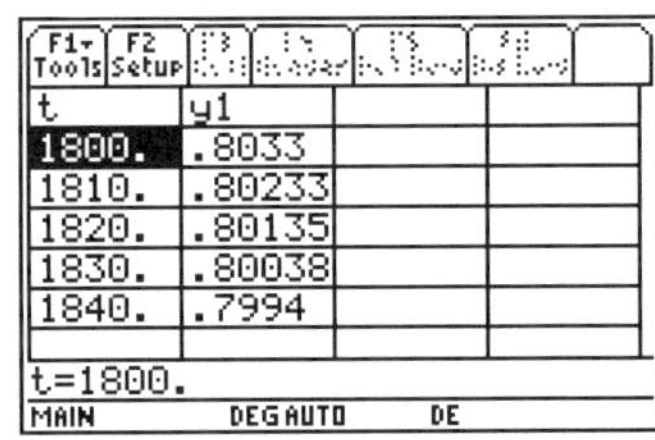

t	y1
1800.	.8033
1810.	.80233
1820.	.80135
1830.	.80038
1840.	.7994

t=1800.

MAIN DEG AUTO DE

Example 4: Higher-order differential equations

In this example, you will solve a second-order differential equation. Since the TI-89 only accepts first-order differential equations, you will need to use a set of substitutions to solve higher-order equations. These substitutions are:

$$y1' = y2$$

$$y1'' = y2'$$

When you make these substitutions, you can enter the second-order differential equation in $y2' =$.

Solve

$$y'' = -4y, \quad y(0) = -3, \quad y'(0) = 0$$

Solution

First, graph the solution to the equation. Then use the **deSolve(** command to find an analytic solution.

1. In the Y= Editor, use the substitutions described above and enter the equations and initial conditions shown at the right. Deselect $y2'$ by moving the cursor to $y2'$ line and pressing [F4] ✓.

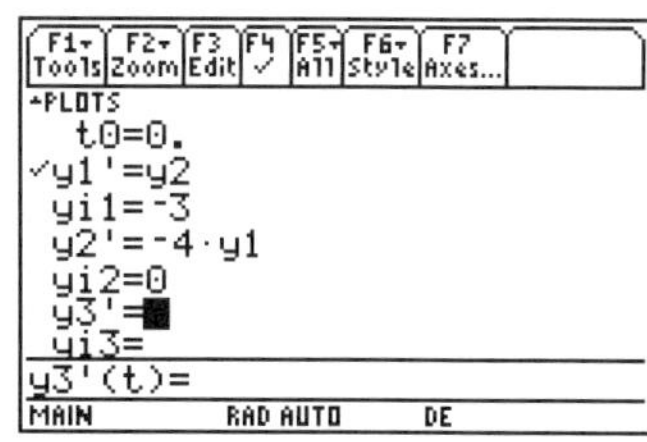

2. In the Window Editor, enter the Window values as shown at the right.

3. Press [F1] **Tools**, and select **9:Format**, and make sure **Solution Method = RK** and **Fields = FLDOFF**.

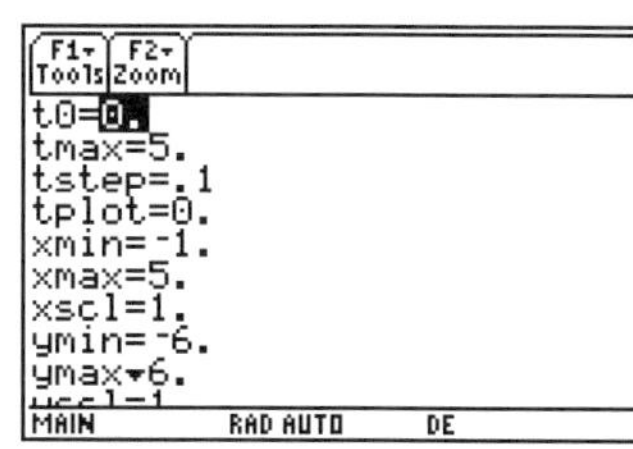

4. Press [◆] [GRAPH] to graph the solution.

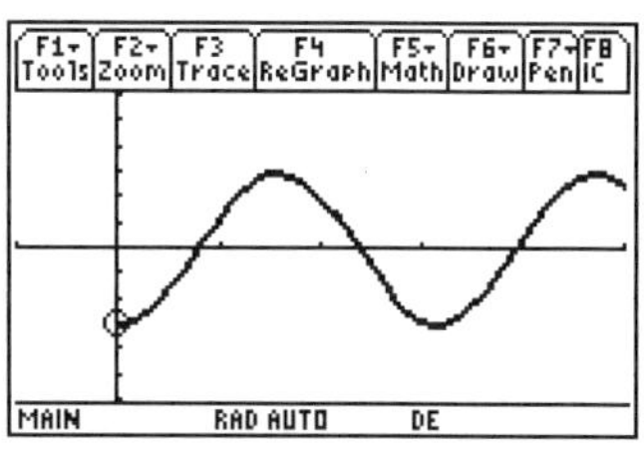

5. Now enter the **deSolve(** command on the Home screen to find an analytic solution.

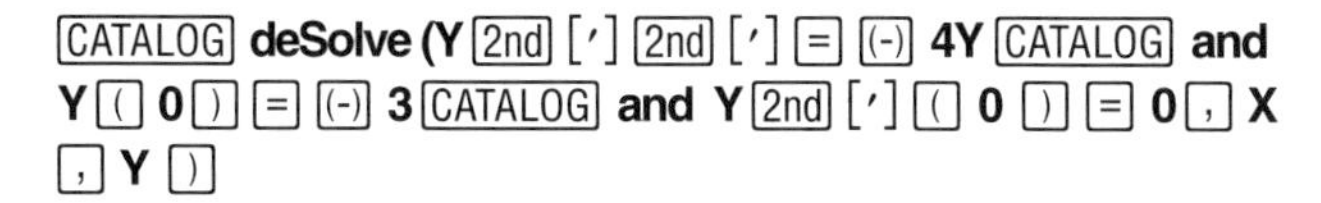

[CATALOG] **deSolve (Y** [2nd] [′] [2nd] [′] [=] [(-)] **4Y** [CATALOG] **and Y** [(] **0** [)] [=] [(-)] **3** [CATALOG] **and Y** [2nd] [′] [(] **0** [)] [=] **0** [,] **X** [,] **Y** [)]

The analytic solution is $y = -3\cos(2x)$.

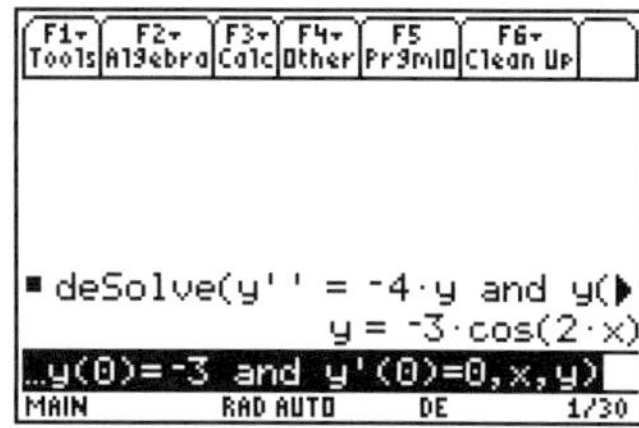

5. Use the **DrawFunc** command to compare the graph of the analytic solution with the graph obtained in differential equation graphing mode.

   [CATALOG] **DrawFunc** [(-)] **3** [×] [2nd] [COS] **2X** [)] [ENTER]

6. Press [◆] [GRAPH]. The graph of $y = -3\cos(2x)$ enters the left part of the screen and continues over the graph produced by the differential equation grapher. This is evidence that you have the right solution.

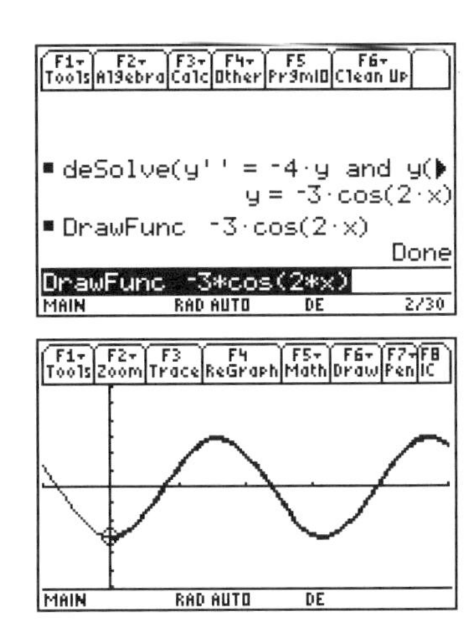

Example 5: Modeling baseball trajectories

This example shows how to change the axes in differential equation graphing mode to model the path of a projectile.

A baseball is hit when it is 3 feet above the ground. It leaves the bat with an initial velocity of 152 ft/sec at an angle of 20° with the horizontal. How far will the ball travel?

Solution

The baseball moves both horizontally and vertically. Assume that the horizontal deceleration due to air resistance is directly proportional to the horizontal velocity. The vertical acceleration due to gravity is -32 ft/s^2, and the vertical deceleration due to air resistance is proportional to the vertical velocity. Assume a constant of proportionality for air resistance of -0.05. Make the following substitutions in order to enter this problem on the TI-89:

horizontal distance = $x = y1$

initial horizontal distance = $yi1 = 0$

horizontal velocity = $y1' = y2$

initial horizontal velocity = $yi2 = 152\cos(20°)$

horizontal acceleration = $y1'' = y2' = -.05y2$

vertical distance = $y = y3$

initial vertical distance = $yi3 = 3$

vertical velocity = $y3' = y4$

initial vertical velocity = $yi4 = 152\sin(20°)$

vertical acceleration = $y4' = -32 - .05y4$

1. In the MODE dialog box, set **Angle=DEGREE**.

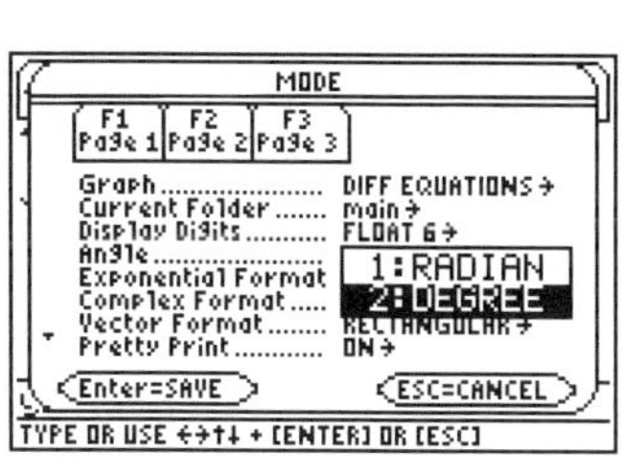

2. In the Y= Editor, enter the equations and initial conditions.

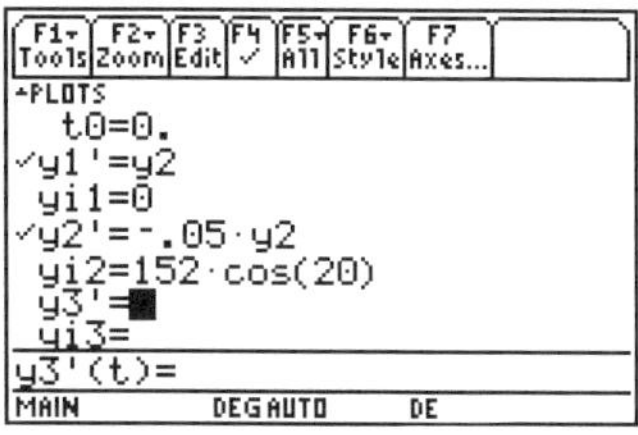

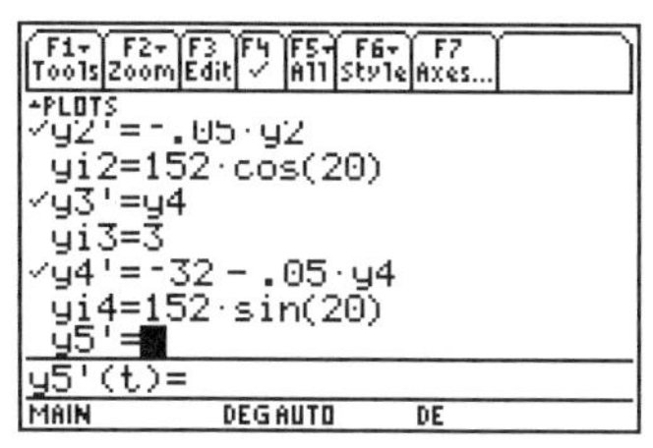

3. Press [F1] **Tools** and select **9:Format**. Set **Solution Method = RK**. Press [ENTER].

4. Specify the x- and y- axes by pressing [2nd] [F7] **Axes**. In the AXES dialog box, set the following:

   Axes= CUSTOM
   x Axis= y1
   yAxis= y3

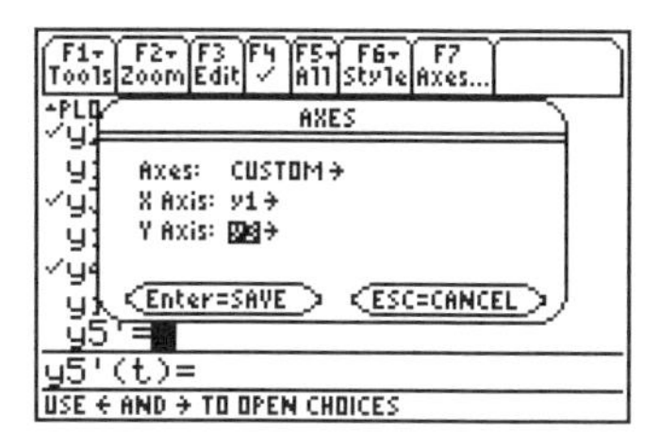

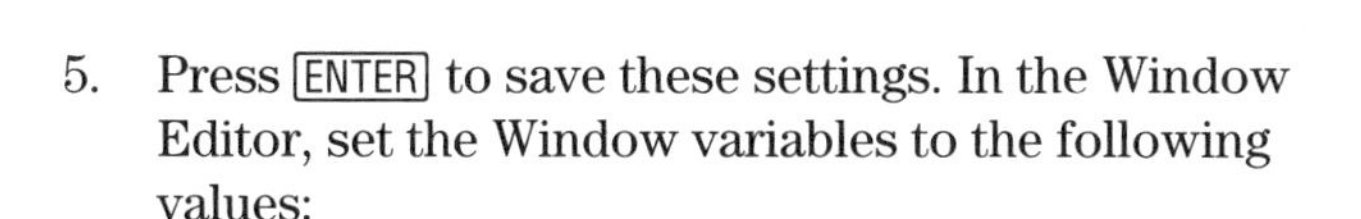

5. Press [ENTER] to save these settings. In the Window Editor, set the Window variables to the following values:

t0 =0	**xscl =50**
tmax =5	**ymin = -15**
tstep =.1	**ymax =100**
tplot =0	**yscl =10**
xmin =0	**ncurves =0**
xmax =500	**diftol =.001**

6. Press [♦] [GRAPH] to graph the solution. The path should resemble the trajectory of a baseball. Press [F3] **Trace** to find the horizontal distance traveled by the ball.

   The ball travels about 422 feet.

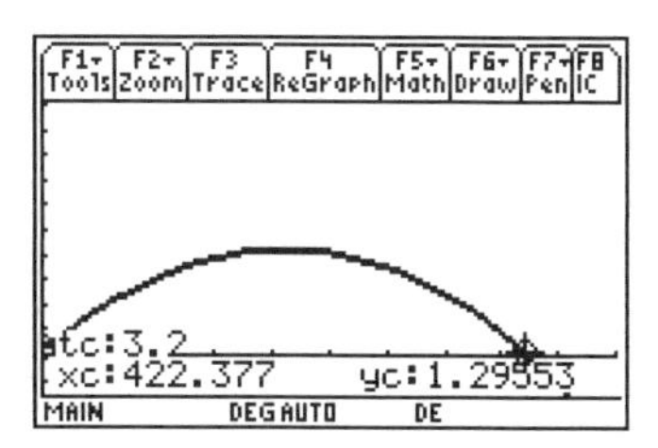

Exercises

1. Graph the slope field for the differential equation

$$\frac{dy}{dx} = \cos(x)$$

2. Graph the solution for the differential equation

$$\frac{dy}{dx} = \cos(x)$$

with initial condition $y(0)=0$. Graph the specific solution first with the slope field and then without it.

3. Graph the slope field for the differential equation

$$\frac{dy}{dx} = \frac{\sin(x)}{x}$$

4. Graph the solution for the differential equation

$$\frac{dy}{dx} = \frac{\sin(x)}{x}$$

with initial condition $y(-7)=-1$. Graph the specific solution first with the slope field and then without it.

Solve exercises 5 and 6 using tabular zoom.

5. How old is a sample in which 50% of the carbon-14 has decayed?
6. How old is a sample in which 80% of the carbon-14 has decayed?

Graph the solutions to the following higher-order differential equations (exercises 7 and 8) and where possible find the analytic solution. Compare the graph of the analytic solution with the graph made by the TI-89 in differential equation graphing mode.

7. $y'' = -9.8, \quad y(0) = 0, \quad y'(0) = 41$
8. $y'' = 1 - 2y - 2y', \quad y(0) = 0, \quad y'(0) = 0$
9. Find the horizontal distance traveled by a baseball hit when it is 3 feet above the ground. The baseball leaves the bat with an initial velocity of 121 ft/sec at an angle of 36° with the horizontal. Assume the air resistance constant is -0.05.
10. If there is a 10-foot fence 375 feet away in the outfield, is the ball in Exercise 9 a home run?

Chapter 8

Parametric, Vector, Polar, and 3D Functions

In this chapter, you will graph parametric functions, vectors, polar functions, and 3D functions on the TI-89.

Parametric functions

The motion of a particle moving in a plane can often be described by assuming the x- and y- coordinates are both functions of time. For example, $x = \cos(t)$, $y = \sin(t)$. Since x and y both have a common parameter t, these are called *parametric equations*. You can graph parametric functions on the TI-89 setting **Graph=PARAMETRIC** in the MODE dialog box. Also set **Angle=RADIAN**.

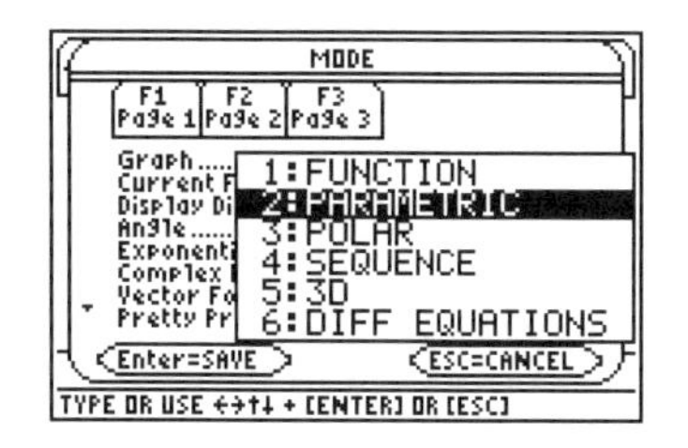

Example 1: Parametric equations for a circle

Graph the parametric equations $x = \cos(t), \quad y = \sin(t)$.

Solution

1. In the Y= Editor, clear all equations. Then enter the equations above in $xt1$ and $yt1$.

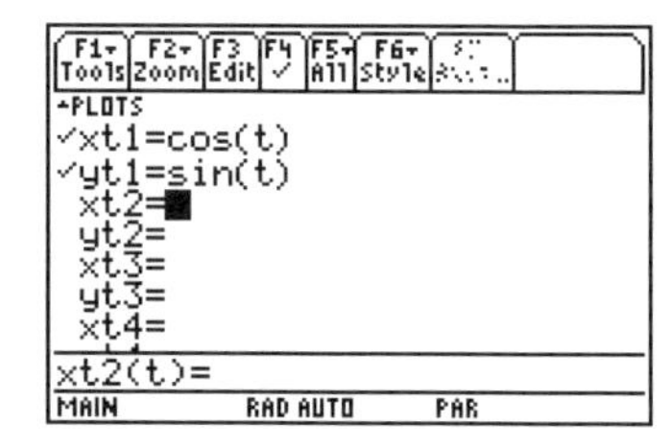

2. Press [◆] [WINDOW] to set the viewing window. The Window variable **tmin** (0) is the starting value for t and **tmax** (2π) is the final value. The increment in t from one point to the next in the graph is given by **tstep** ($\pi/24$). Enter the values for the Window variables as shown.

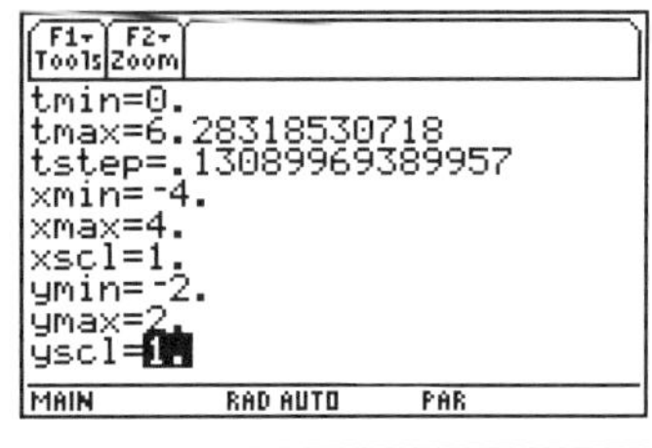

3. Press [◆] [GRAPH] to graph the parametric equations.

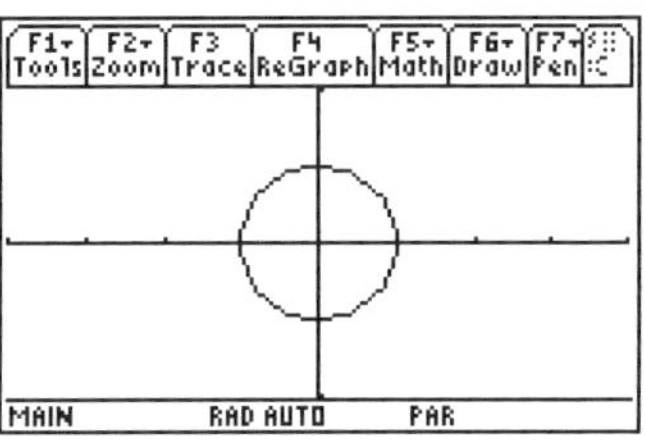

Example 2: Slope of a parametric curve, chain rule for parametric equations

Find the slope of the circle in Example 1 at $t = \dfrac{\pi}{4}$.

Solution

The slope of the curve is given by $\dfrac{dy}{dx}$. You can approximate this value from the MATH menu on the Graph screen.

1. With the graph from Example 1 displayed, press [F5] **Math** and select **6:Derivatives**.

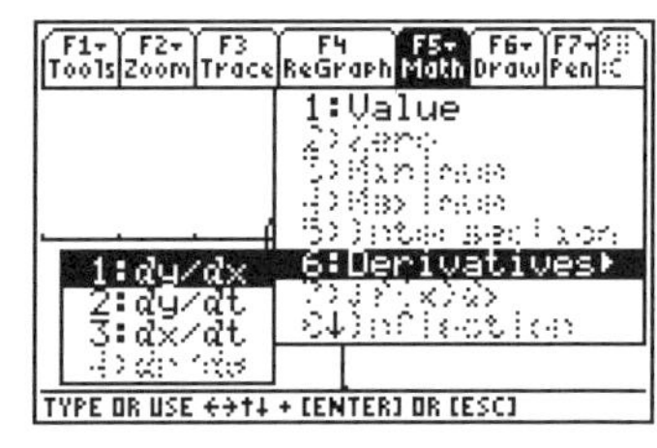

2. Select **1:dy/dx**. The TI-89 returns to the graph and prompts for a t-coordinate.

3. Enter the value.

   [2nd] [π] [÷] **4**

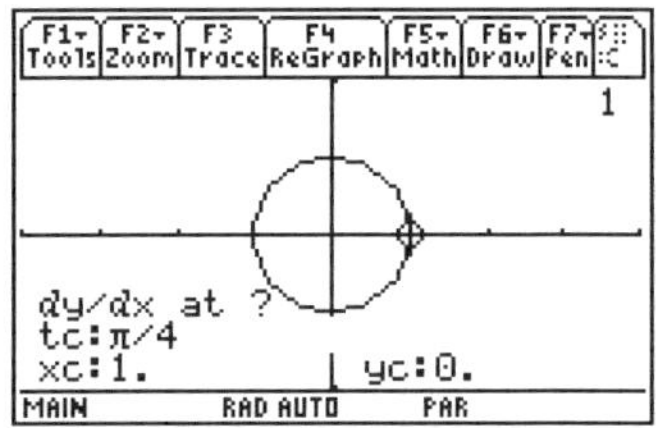

4. Press [ENTER].

   The slope of the curve is -1.

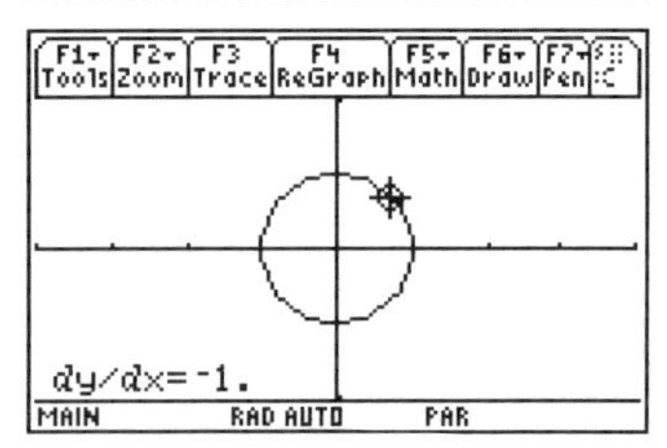

You can verify this result with the chain rule for parametric equations, which says that

$$\frac{dy}{dx} = \frac{\frac{dy}{dt}}{\frac{dx}{dt}}$$

5. Return to the Home screen and define a function that gives the slope by entering:

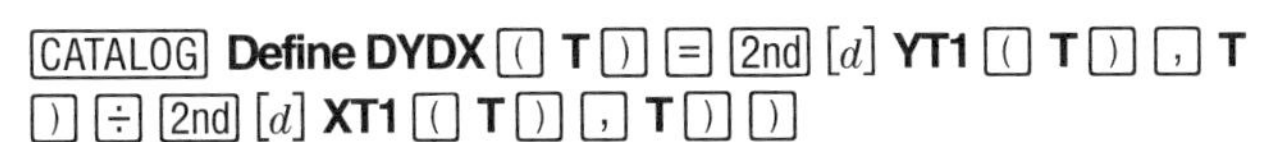

[CATALOG] **Define DYDX** [(] **T** [)] [=] [2nd] [d] **YT1** [(] **T** [)] [,] **T** [)] [÷] [2nd] [d] **XT1** [(] **T** [)] [,] **T** [)] [)]

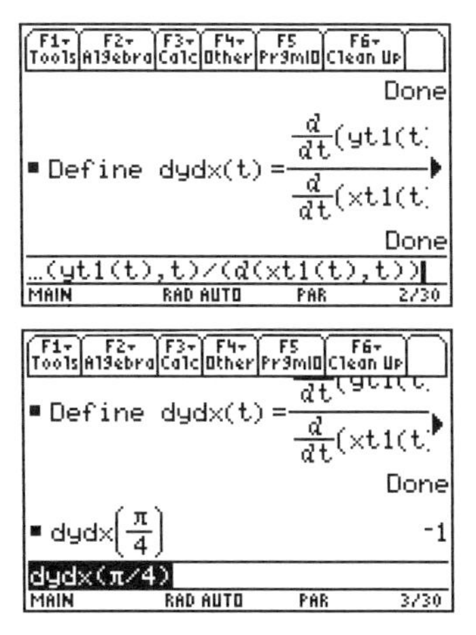

6. Evaluate this function at $t = \frac{\pi}{4}$.

   The slope function you defined gives the same result as that returned by the GRAPH MATH dy/dx feature.

Example 3: Arc length of parametric curves

This example defines a function to calculate the arc length of a parametric curve.

Find the length of one arch of the cycloid x=t−sin(t), y=1−cos(t).

Solution

Arc length is given by the definite integral

$$\int_a^b \sqrt{\left(\frac{dx}{dt}\right)^2 + \left(\frac{dy}{dt}\right)^2}\, dt$$

1. Press [2nd] [F6] **Clean Up** and select **2:NewProb** to clear variables and set other defaults.

2. Define the above integral with the command **Define alength**.

   [CATALOG] **Define ALENGTH** [(] **T** [,] **A** [,] **B** [)] [=] [2nd] [∫] [2nd] [√] [(] [2nd] [d] **XT1** [(] **T** [)] [,] [)] [^] **2** [+] [(] [2nd] [d] **YT1** [(] **T** [)] [,] **T** [)] [)] [^] **2** [)] [,] **T** [,] **A** [,] **B** [)] [ENTER]

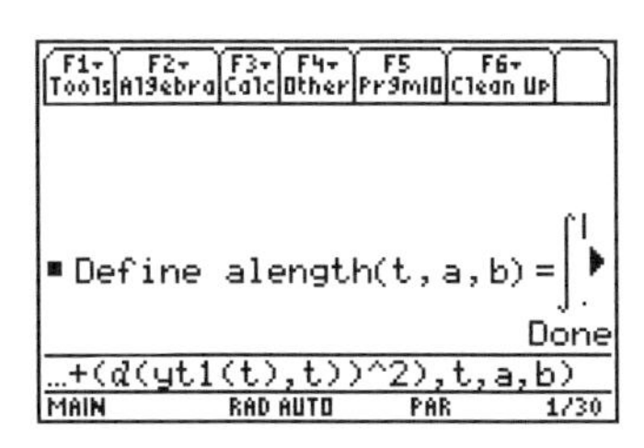

3. In the Y= Editor, enter the parametric equations for the cycloid.

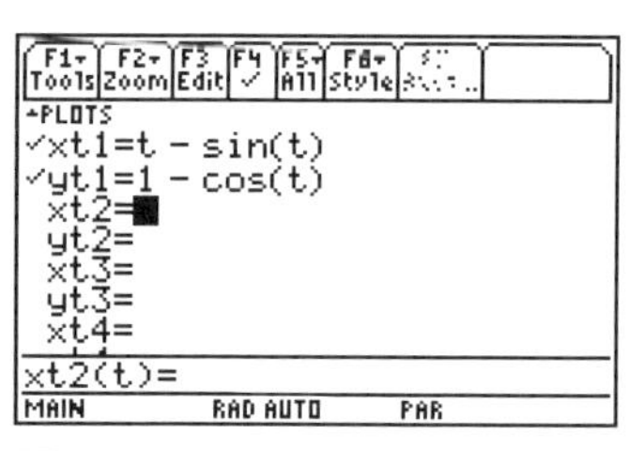

4. In the Window Editor, set the values as shown for a [0,2π] x [-1,3] window with **xscl** = π.

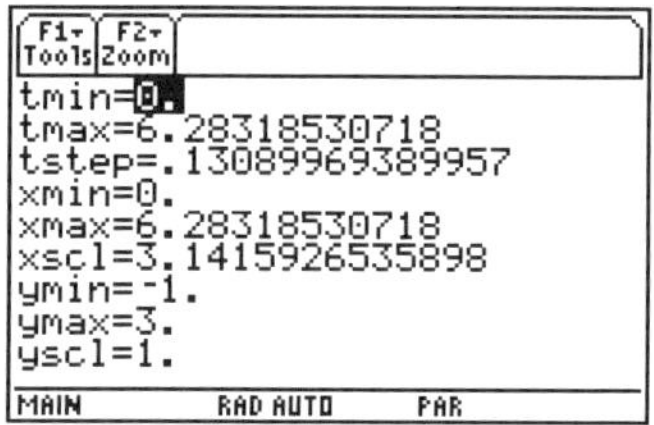

5. Press ◆ [GRAPH] to graph the arch.

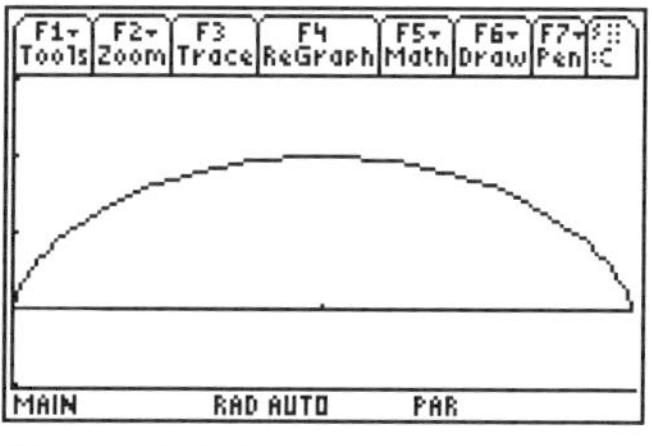

6. Since one arch is completed for $0 \le t \le 2\pi$, return to the Home screen and enter the command **alength(t,0,2π)** to find the arc length of the arch.

   The length of the arch is 8.

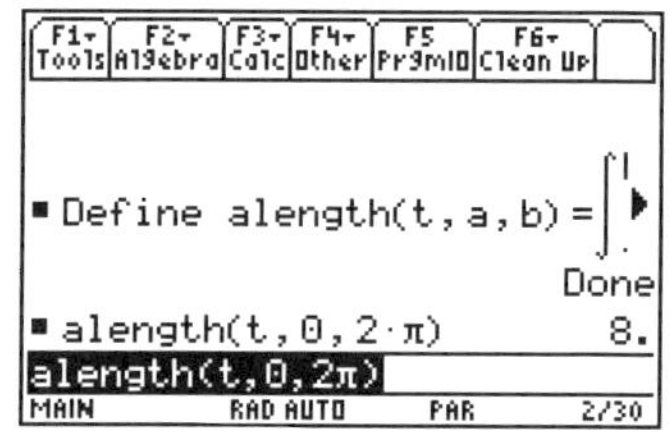

Example 4: Parametric equations for trajectories

In this example, you will revisit the baseball problem from Chapter 7, Example 5. This time the goal is to find the parametric equations for the position of the baseball from the information given about acceleration.

A baseball is hit when it is 3 feet above the ground. It leaves the bat with an initial velocity of 152 ft/sec at an angle of 20° with the horizontal. How far will the ball travel?

Solution

The baseball moves both horizontally and vertically. Assume that the horizontal deceleration due to air resistance is directly proportional to the horizontal velocity. The vertical acceleration due to gravity is -32 ft/s^2, and the vertical deceleration due to air resistance is proportional to the vertical velocity. Assume a constant of proportionality for air resistance of -0.05.

You can find the parametric equations for the position of the baseball by solving the differential equations

$x'' = -.05x'$, $x(0)=0$, $x'(0)=152\cos(20°)$

$y'' = -32 - .05y'$, $y(0)=3$, $y'(0)=152\sin(20°)$

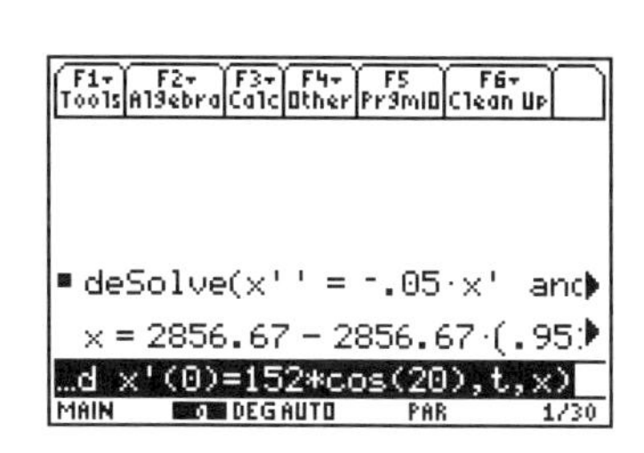

1. Press [2nd] [F6] **Clean Up** and select **2:NewProb** to clear variables and set other defaults. In the MODE dialog box, set **ANGLE=DEGREE**.

2. Solve the first differential equation with the command **deSolve(x″= -.05x′ and x(0)=0 and x′(0)=152cos(20),t,x)**.

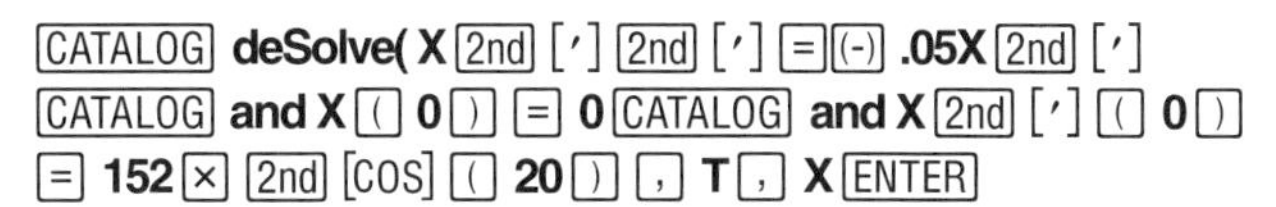

The solution is x=2856.67−2856.67(.951229)^t.

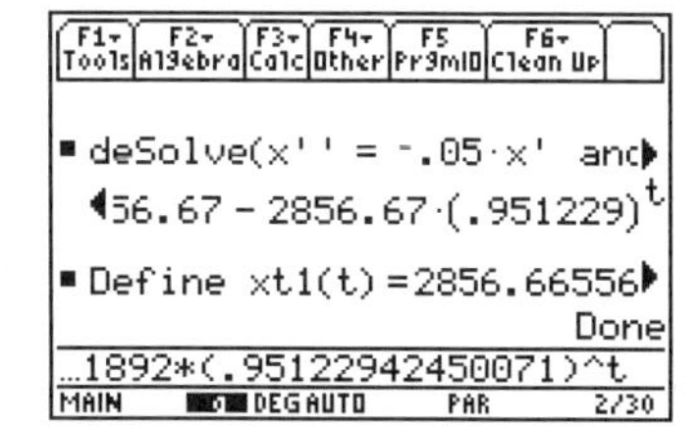

3. Store this in $xt1$ with the **Define** command.

   [CATALOG] **Define XT1** [(] **T** [)] [=] [▲] [ENTER]

   Press [◄] to move left and then press [←] to delete **x=**.

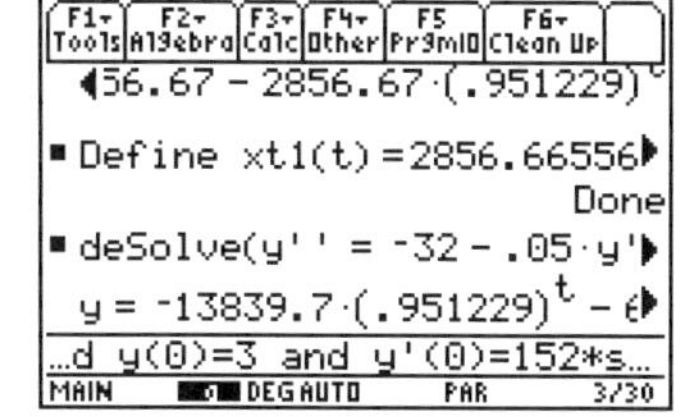

4. Solve the second differential equation with the command **deSolve(y″= -32 -.05y′ and y(0)=3 and y′(0)=152sin(20),t,y)**.

   The solution is y=−13839.7(.951229)^t−640t+13842.7

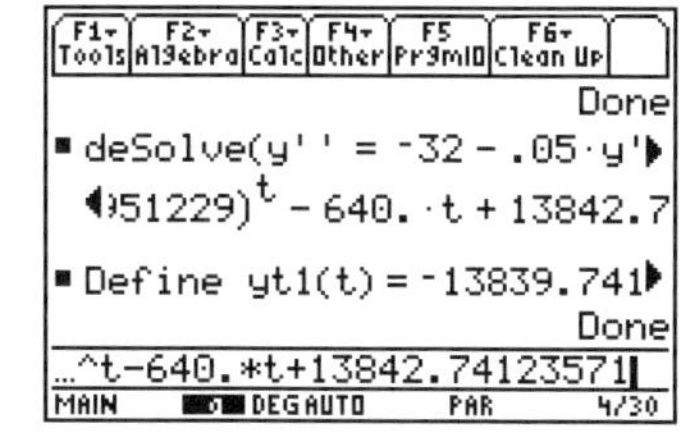

5. Store this solution in $yt1$ with the command **Define yt1(t)=-13839.7(.951229)^t-640t+13842.7**. Paste the result of the last command to simplify entering this command.

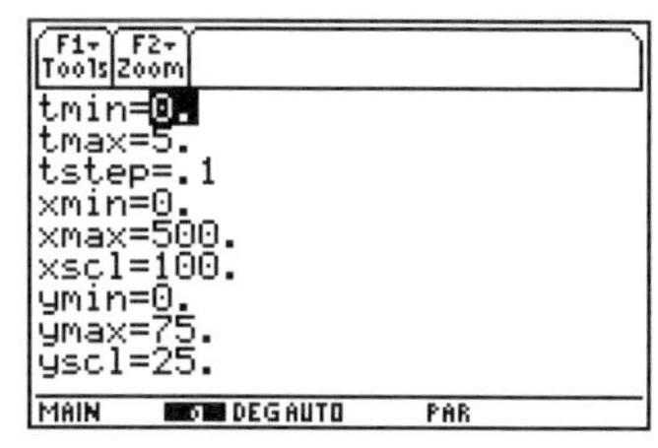

6. In the Window Editor, set up the viewing window as shown.

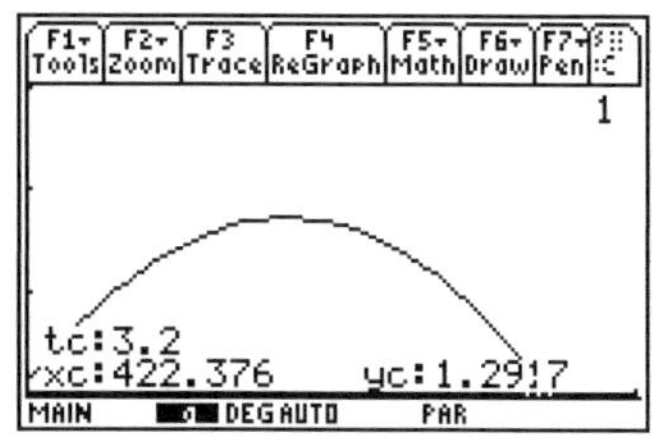

7. Graph the parametric equations and trace to see how far the baseball traveled.

   The ball travels about 422 feet.

Vectors

If $\boldsymbol{r} = x(t)\boldsymbol{i} + y(t)\boldsymbol{j}$ is the position vector for a projectile, the velocity vector is given by $\boldsymbol{v} = x'(t)\boldsymbol{i} + y'(t)\boldsymbol{j}$ and the acceleration vector is $\boldsymbol{a} = x''(t)\boldsymbol{i} + y''(t)\boldsymbol{j}$. In this example, one-dimensional matrices are used to represent vectors on the TI-89.

Example 5: Symbolic and graphical representation of vectors

Represent the position vector

$$\boldsymbol{r}(\text{t}) = \frac{4}{5}\cos(2\text{t})\boldsymbol{i} + \frac{3}{2}\sin(2\text{t})\boldsymbol{j}$$

and the velocity and acceleration vectors symbolically and graphically.

Solution

1. Press [2nd] [F6] **Clean Up** and select **2:NewProb** to clear variables and set other defaults. In the MODE dialog box, set **Angle=RADIAN**.

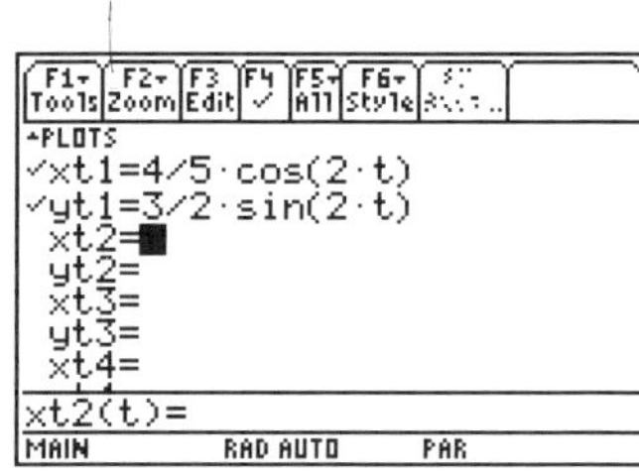

2. In the Y = Editor, enter the parametric equations that correspond to the position vector as $xt1$ and $yt1$.

3. In the Window Editor, set up the viewing window as shown.

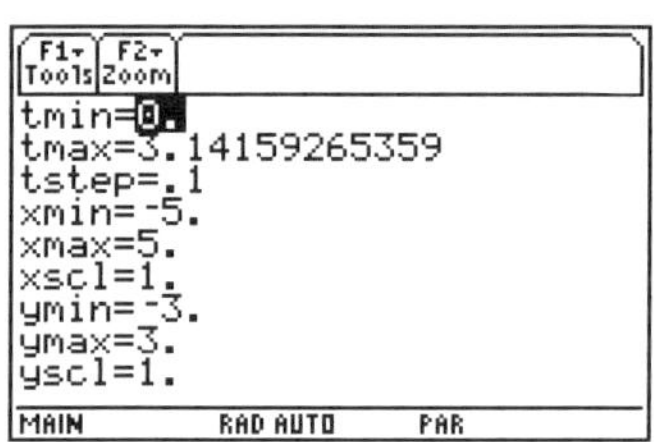

4. Return to the Home screen and define the *position vector* function **r**(*t*) with the command **Define r(t)=[xt1(t),yt1(t)]**.

   [CATALOG] **Define R** [(] **T** [)] [=] [2nd] [[] **XT1** [(] **T** [)] [,] [(] **YT1** [(] **T** [)] [2nd] []] [ENTER]

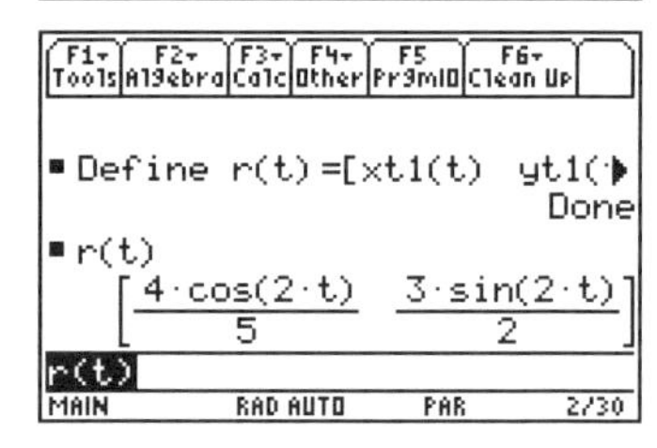

5. Display $\boldsymbol{r}(t)$.

   R [(] **T** [)] [ENTER]

6. Define the *velocity vector* as the first derivative of the position vector with the command **Define v(t)=*d*(r(t),t)** and display $\boldsymbol{v}(t)$.

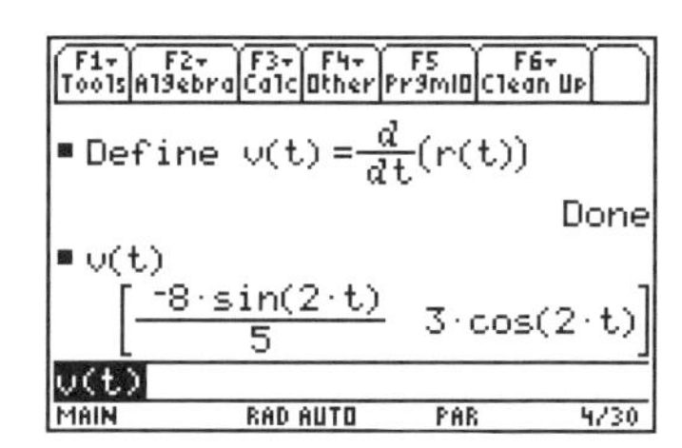

7. Define the *acceleration vector* as the second derivative of the position vector with the command **Define a(t)=** d**(r(t),t,2)**, and display $\boldsymbol{a}(t)$.

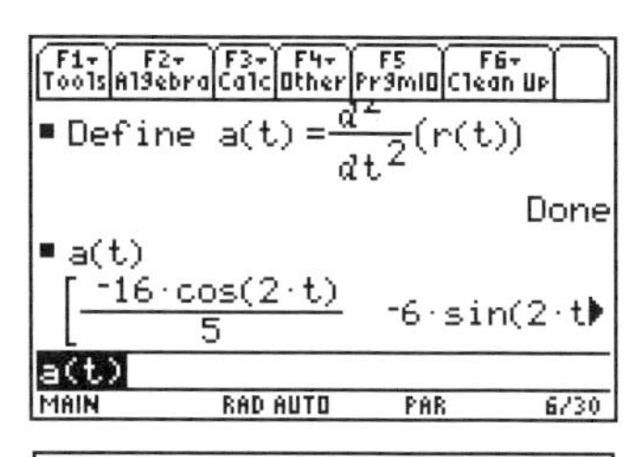

8. Now enter the following commands to graph each of the vectors for $t = .6$. You can enter the **Line** command one letter at a time or copy it from the CATALOG.

```
.6→c
r(c)[1,1]→p
r(c)[1,2]→q
Line 0,0,p,q
Line p,q,p+v(c)[1,1],q+v(c)[1,2]
Line p,q,p+a(c)[1,1],q+a(c)[1,2]
```

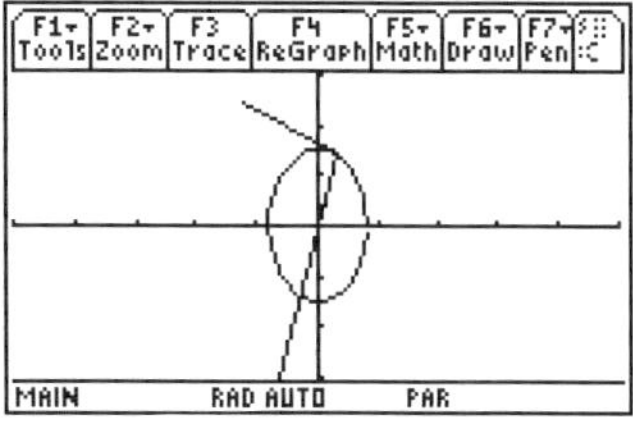

Polar functions

You must change the graphing mode to **POLAR** to graph polar functions.

Example 6: Area bounded by polar curves

Use polar graphing to find the area of the region that lies inside the circle $r = 1$ and outside the cardioid $r = 1 - \cos\theta$.

Solution

1. Press [2nd] [F6] **Clean Up** and select **2:NewProb** to clear variables and set other defaults. In the MODE dialog box, set **GRAPH=POLAR**.

2. Press [♦] [Y=] to display the Y= Editor. Then enter the equations in $r1$ and $r2$. Press [♦] [^] to enter θ.

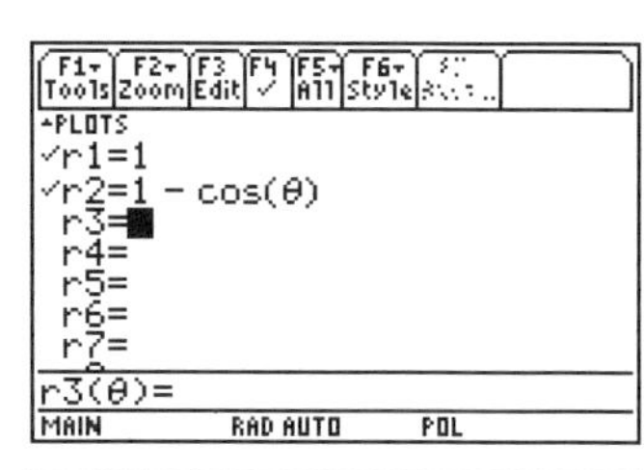

3. In the Window Editor, set up the viewing window as shown.

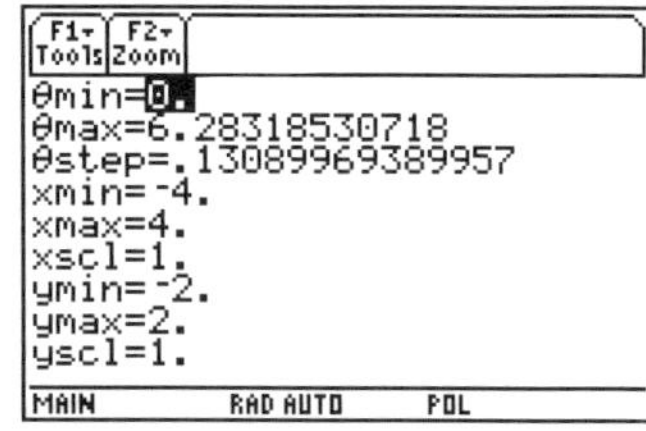

4. Graph the polar curves.

   From the graph, it appears that the polar functions intersect at $\theta = -\frac{\pi}{2}$ and $\frac{\pi}{2}$.

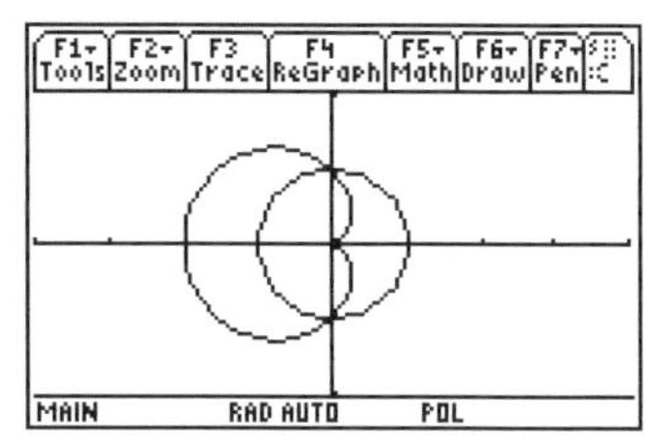

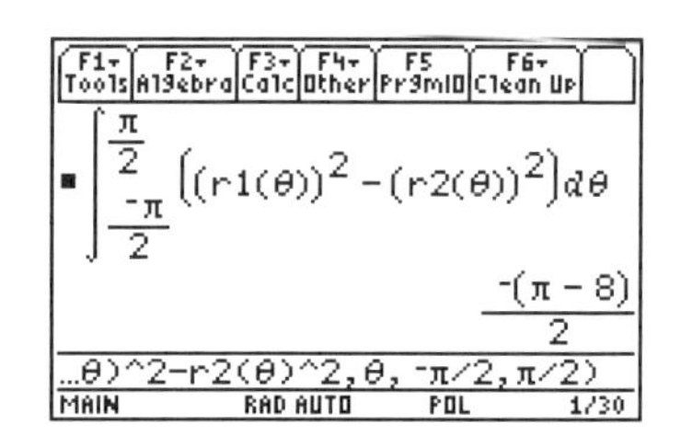

5. The area is given by entering the definite integral ∫ **(r1(θ)^2- r2(θ)^2,θ,-π/2, π/2)** on the Home screen and then dividing the result by 2.

   [2nd] [∫] **R1** [(] [◆] [θ] [)] [^] **2** [-] **R2** [(] [◆] [θ] [)] [^] **2** [,] [◆] [θ] [,] [(-)] [2nd] [π] [÷] **2** [,] [2nd] [π] [÷] **2** [)] [ENTER]

   The area is

$$\frac{8-\pi}{4}$$

3D functions

To graph functions of two variables, select 3D in the MODE dialog box.

Example 7: The graph of a saddle

Using 3D graphing, graph the "saddle" $z = x^2 - y^2$ in a [-2,2] x [-2,2] x [-2,2] viewing window.

Solution

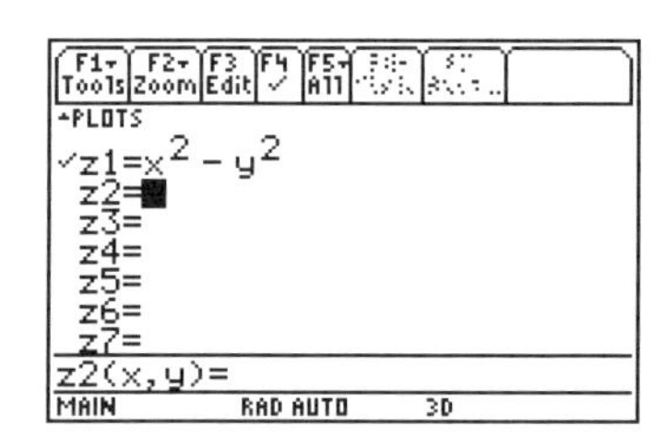

1. Press [2nd] [F6] **Clean Up** and select **2:NewProb** to clear variables and set other defaults. In the MODE dialog box, set **GRAPH=3D**.
2. Press [◆] [Y=] to display the Y= Editor. Enter the equation in $z1$.
3. Press [◆] [WINDOW] and enter the following window values:

eyeθ= 30	**ymin= -2**
eyeφ= 70	**ymax= 2**
eyeψ= 0	**ygrid= 14**
xmin= -2	**zmin= -2**
xmax= 2	**zmax= 2**
xgrid= 14	**ncontour= 5**

4. Press [◆] [GRAPH] to see the surface.

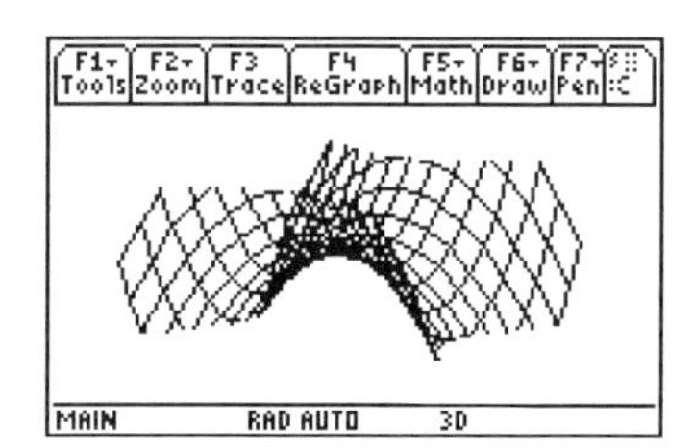

5. You can rotate the surface with the cursor movement keys.

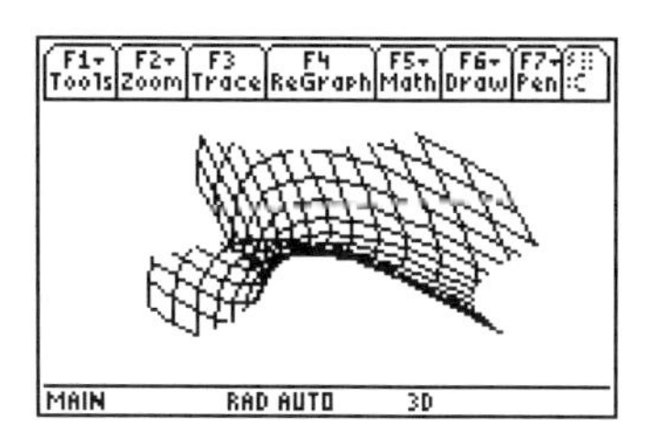

6. Return to the Window Editor to see how the rotations affect the viewing angle variables (**eyeθ**, **eyeϕ**, and **eyeψ**, for example).

Exercises

1. Given the parametric equations $x = 3\cos(t)$ and $y = 2\sin(t)$ for $0 \le t \le 2\pi$:

   (a) Graph the curve described by these parametric equations.
   (Use viewing window [-3,3] x [-6,6]).

   (b) Use the **Math** menu on the Graph screen to find the slope of the parametric curve at

   $$t = \frac{\pi}{4}.$$

   (c) Use the chain rule for parametric equations to find the slope of the parametric curve at $t = \frac{\pi}{4}$.

   (d) Find the length of the parametric curve.

2. Given the parametric equations $x = \sin(2t)$ and $y = \sin(t)$ for $0 \le t \le 2\pi$.

   (a) Graph the curve described by these parametric equations.
   (Use viewing window [-4,4] x [-2,2]).

   (b) Use the **Math** menu on the Graph screen to find the slope of the parametric curve at
   $t = \frac{\pi}{2}$.

   (c) Use the chain rule for parametric equations to find the slope of the parametric curve at. $t = \frac{\pi}{2}$.

   (d) Find the length of the parametric curve.

3. A baseball is hit when it is 3 feet above the ground. It leaves the bat with an initial velocity of 127 ft/sec at an angle of 30° with the horizontal. The baseball moves both horizontally and vertically. Assume that the horizontal deceleration due to air resistance is directly proportional to the horizontal velocity. The vertical acceleration due to gravity is 32 ft/s^2 and the vertical deceleration due to air resistance is proportional to the vertical velocity. Assume a constant of proportionality for air resistance of -0.05.

   Find the parametric equations for the motion of the ball, and use the parametric graph to determine how far will the ball travel.

4. Represent the position vector $r = 4\cos(t)\, i + 3\sin(t)\, j$ and the velocity and acceleration vectors at t=0.6 symbolically and graphically. (Use viewing window [-10,10] x [-5,5].)

5. Use polar graphing to find the area inside the circle $r = 2\sin(\theta)$ and outside the cardioid $r = 1 - \cos(\theta)$.

6. Graph the following three-dimensional surfaces.

   (a) $z = 9 - x^2 - y^2$

   (b) $z = x * \sin(y)$

   (c) $z = \cos(x * y)$

Chapter 9

Infinite Sequences and Series

An *infinite sequence* is a list or expression of the form: a_1, a_2, a_3, ... , a_n. An infinite sequence could be defined as a function with a domain consisting of the counting numbers. Since a sequence is a function, it can be represented symbolically, numerically, and graphically.

Before you begin

Because the domain of a sequence consists of only the counting numbers, a different graphing mode is used on the TI-89 called the Sequence graphing mode. Press [MODE] and set **Graph= SEQUENCE** before you complete the first two examples in this chapter.

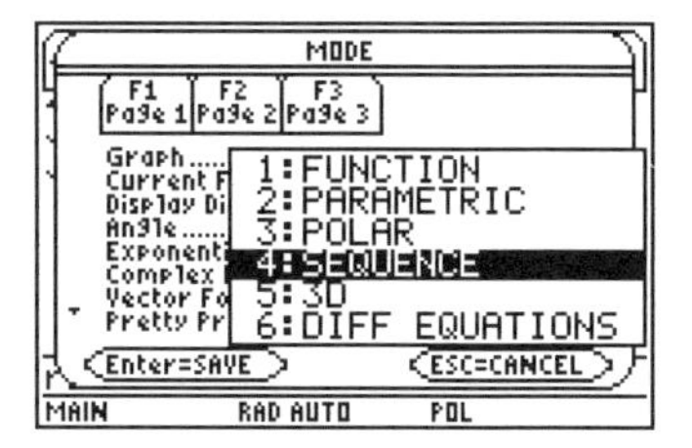

Example 1: Investigating sequences numerically, graphically, symbolically

Investigate the behavior of the sequence

$$a_n = \frac{2^n}{n!}$$

for large values of n.

Solution

First, user the **seq** (sequence) command to look at the sequence numerically. Then graph the sequence to examine its behavior. Finally, use the limit(command to evaluate the sequence symbolically.

1. Press [2nd] [F6] **Clean Up** and select **2:NewProb** to clear variables and set other defaults.

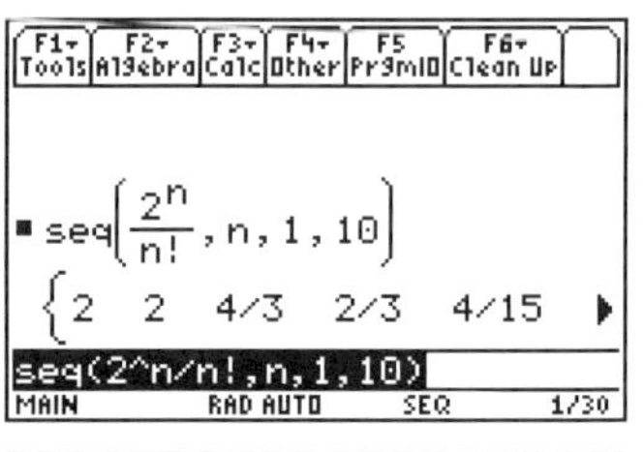

2. Generate several terms of this sequence with the **seq(** command.

   [CATALOG] **seq(2** [^] **N** [÷] **N** [◆] [÷] [,] **N** [,] **1** [,] **10** [)] [ENTER]

3. You can see the rest of the sequence by pressing ▲ to move up to the last line in the history window and then scrolling to the right by pressing ►.

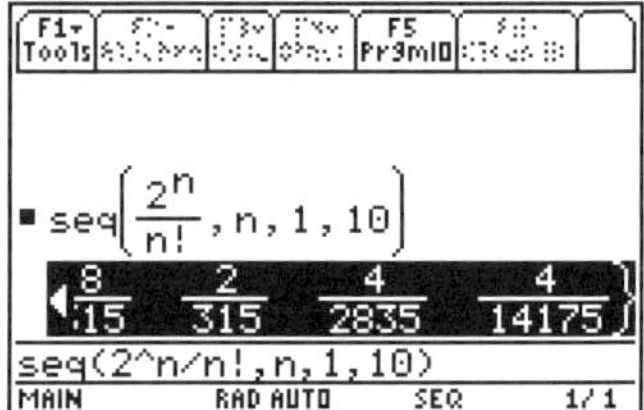

   It appears that the terms in the sequence are approaching 0 as n gets large.

4. Now use a table to represent the sequence. In the Y= Editor, enter the nth term expression for the sequence in $u1$. Clear $ui1$, if necessary.

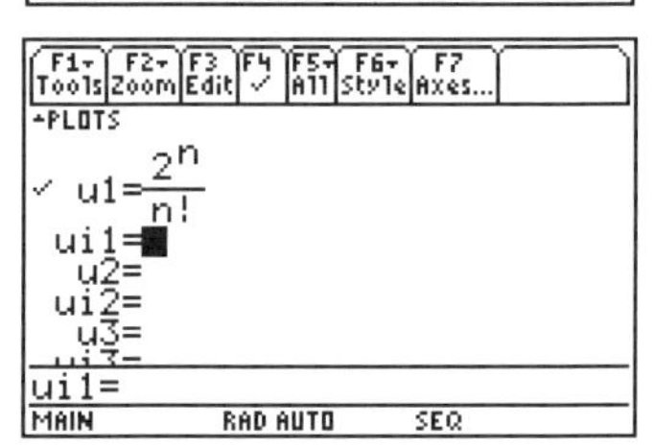

5. Press [◆] [TblSet] to display the TABLE SETUP dialog box, and then enter the values shown.

   The initial value of n is stored in **tblStart** and the increase in n from one row to the next in the table is stored in Δ**tbl**.

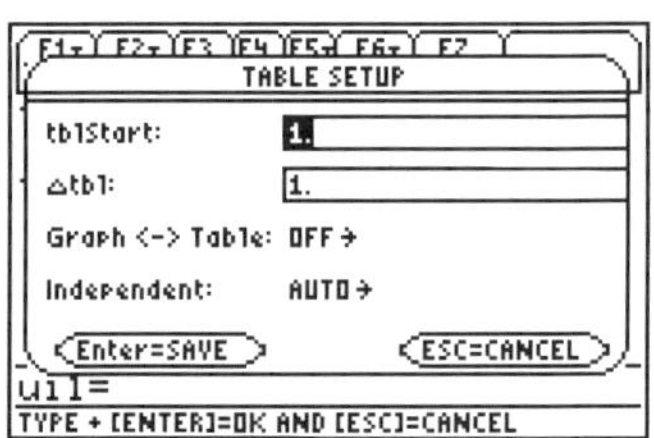

6. After entering these values, press [ENTER]. Then press [◆] [TABLE] to see the table.

n	u1
1.	2.
2.	2.
3.	1.3333
4.	.66667
5.	.26667

n=1.

7. Scroll down the table a row at a time by pressing ▼ or a page at a time by pressing [2nd] ▼.

   By the 15th term, the sequence is very close to zero.

n	u1
11.	.00005
12.	8.6E-6
13.	1.3E-6
14.	1.9E-7
15.	2.5E-8

n=15.

8. Now graph the sequence. Since the nth term is already stored in $u1$, you need only specify a viewing window before graphing the sequence. Press [◆] [WINDOW] and enter the values shown.

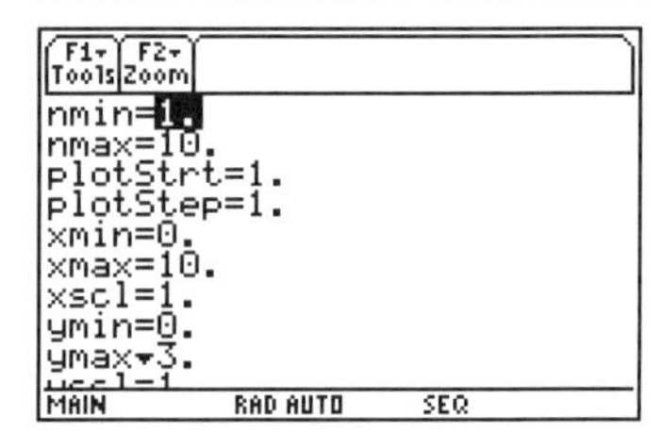

The window variables have the following functions:

Variable	Function
nmin	The minimum or starting value of *n* used to generate the sequence
nmax	The maximum or final value of *n*
plotStrt	The value of *n* at which graphing begins
plotStep	The spacing between plotted points
xmin	The left side of the viewing window
xmax	The right side of the viewing window
xscl	The spacing between the tick marks on the *x*-axis
ymin	The bottom of the viewing window
ymax	The top of the viewing window
yscl	The spacing between the tick marks on the *y*-axis

9. Press [◆] [GRAPH]. The graph appears to converge to the x-axis. This is further evidence the sequence converges to zero.

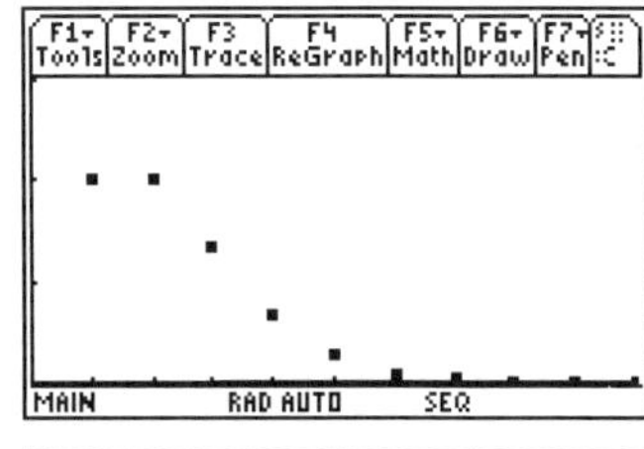

10. Return to the Home screen to confirm this conjecture by evaluating the expression **limit(2^n/n!,n,∞)**.

   [CATALOG] **limit(2** [^] **N** [◆] [÷] [,] **N** [,] [◆] [∞] [)] [ENTER]

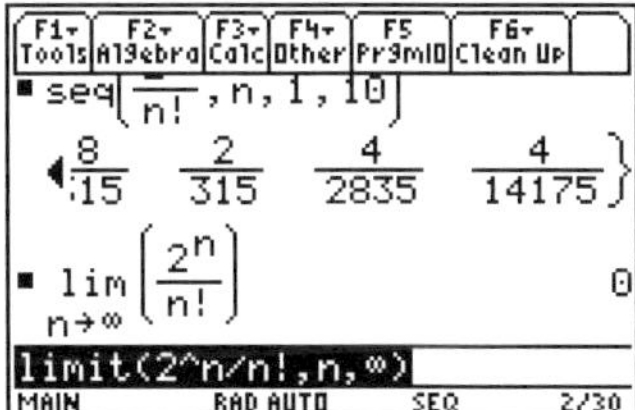

You now have numerical, graphical, and symbolic evidence that the sequence

$$a_n = \frac{2^n}{n!} \text{ converges to zero as } n \text{ gets large.}$$

Example 2: Convergence of an infinite series

An *infinite series* is an expression of the form:

$a_1 + a_2 + a_3 + \ldots + a_k + \ldots$

Another way to represent an infinite series is with the notation:

$$\sum_{k=1}^{\infty} a_k$$

The partial sums of a series are helpful in understanding the behavior of the infinite series. These partial sums form a sequence.

$s_1 = a_1$

$s_2 = a_1 + a_2$

$s_3 = a_1 + a_2 + a_3$

$\vdots$

$s_n = a_1 + a_2 + a_3 + \ldots + a_n = \sum_{k=1}^{n} a_k$

If the sequence of partial sums converges to a limit S as n gets large, we say the infinite series converges to S. If the sequence of partial sums diverges, the infinite series diverges.

Determine if the infinite series

$$\sum_{k=1}^{\infty} \frac{2^k}{k!}$$

converges. If the series converges, estimate its sum.

Solution

You can use the ratio test to establish the convergence of the infinite series. Then investigate the sequence of partial sums graphically and numerically to estimate the sum of the infinite series.

The ratio test says that if

$$\lim_{k\to\infty} \frac{a_{k+1}}{a_k} < 1 \text{, the series } \sum_{k=1}^{\infty} a_k \text{ converges.}$$

1. Press [2nd] [F6] **Clean Up** and select **2:NewProb** to clear variables and set other defaults.

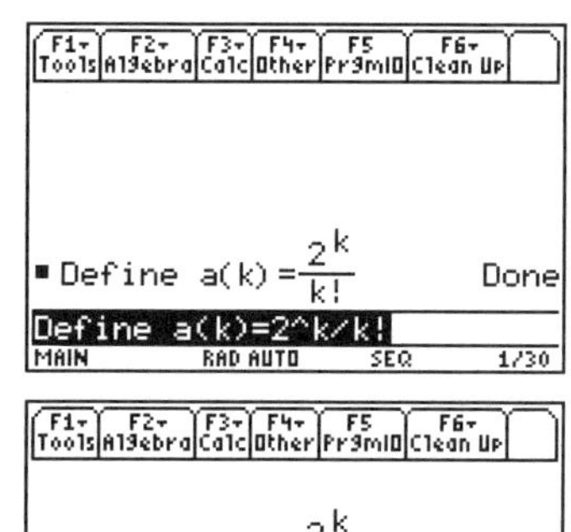

2. Define the kth term in the series with the command **Define a(k)=2^ k/k!**.

   [CATALOG] **Define A** [(] **K** [)] [=] **2** [^] **K** [÷] **K** [♦] [÷] [ENTER]

3. Apply the ratio test.

   [CATALOG] **limit(A** [(] **K** [+] **1** [)] [)] [÷] **A** [(] **K** [)] [,] **K** [,] [♦] [∞] [)] [ENTER]

   Since the limit is less than 1, the infinite series converges. Now estimate the sum of the infinite series.

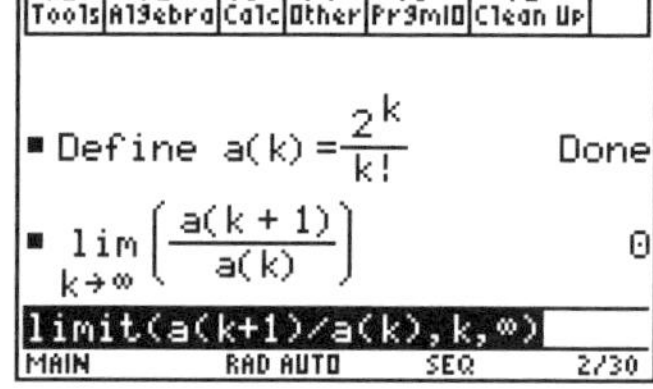

4. In the Y= Editor, enter the expression for the *nth* partial sum, **Σ(2^k/k!,k,1,n)**, in *u*1. To enter **Σ(**, press [2nd] [MATH] and then select **A:Calculus** followed by **4:Σ(sum**.

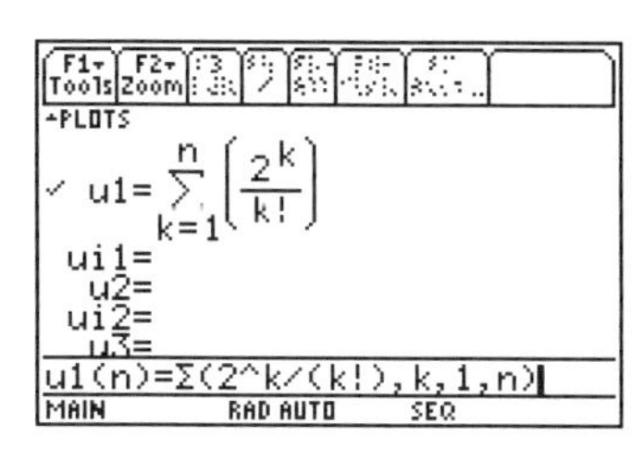

5. Set **tblStart =1** and **Δtbl=1.** In the TABLE SETUP dialog box, press [ENTER] and then press [◆] [TABLE] to display the table.

n	u1
1.	2.
2.	4.
3.	5.3333
4.	6.
5.	6.2667

n=1.

6. As you scroll down the table, you should see a possible limit to the sequence of partial sums.

n	u1
16.	6.3891
17.	6.3891
18.	6.3891
19.	6.3891
20.	6.3891

n=20.

7. Graphing the sequence of partial sums helps to visualize the convergence of the sequence of partial sums.

   In the Window Editor, set up a [0,20] x [0,7] viewing window as shown.

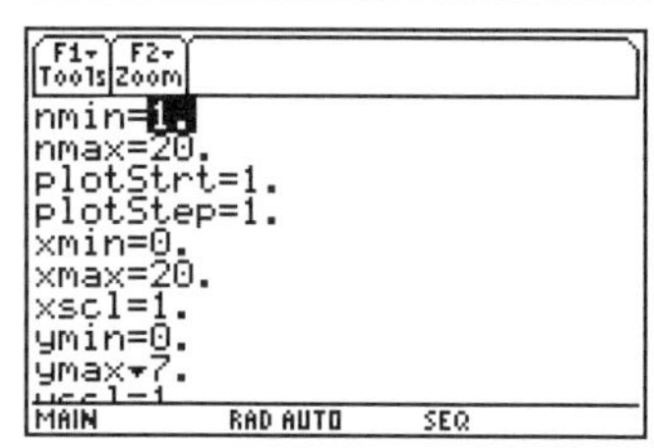

8. Press [◆] [GRAPH] to see the graph.

   The sequence of partial sums appears to level off after 20 terms.

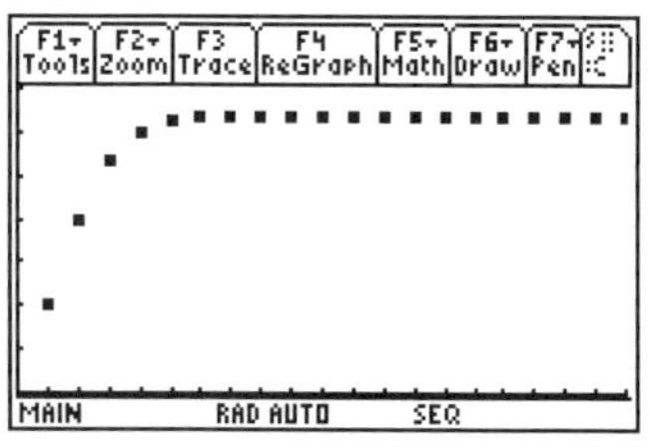

9. Return to the Home screen to find the 50th partial sum by entering **Σ(2^k/k!,k,1,50)**.

   [CATALOG] **Σ(2** [^] **K** [÷] **K** [◆] [÷] [,] **K** [,] **1** [,] **50** [)] [ENTER]

   The result is exact but not very helpful. Press [◆] [ENTER] to repeat the command and obtain a decimal approximation.

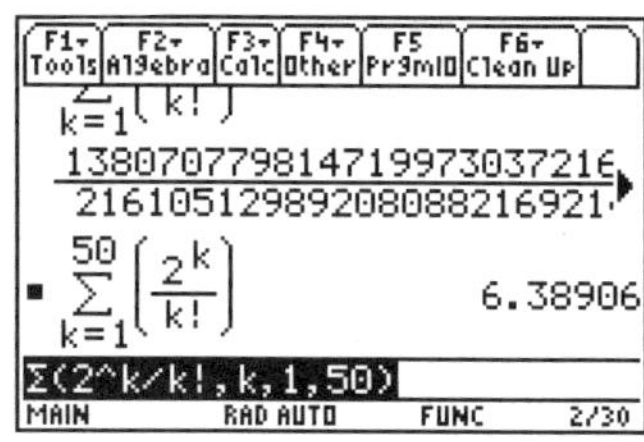

Since the sequence of partial sums appears to converge to approximately 6.389, we estimate that

$$\sum_{k=1}^{\infty} \frac{2^k}{k!} = 6.389.$$

Example 3: Taylor series for $f(x) = e^x$

Taylor series give you polynomials that can be used to approximate other functions. This can be useful when the other function is difficult to evaluate or to manipulate symbolically. You use the fact that if two functions have identical first- and higher-order derivatives at a point, their graphs must be similar.

Determine values for the coefficients of $p(x) = ax^2 + bx + c$ so that the parabola is tangent to the curve $f(x) = e^x$ at $x = 0$.

Solution

The parabola will be tangent to the curve if both functions have the same value at $x = 0$ as well as the same first and second derivatives at $x = 0$.

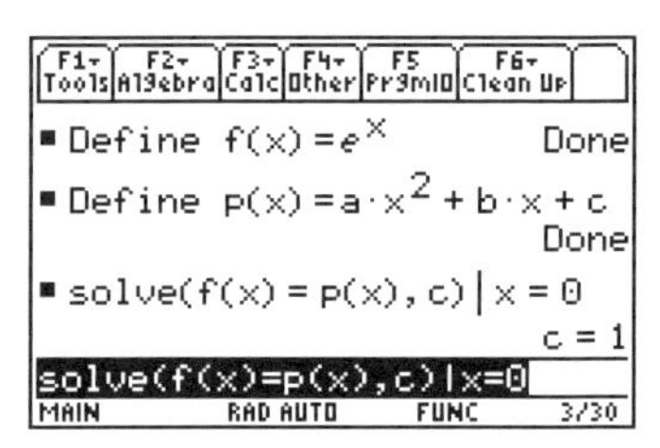

1. Press [2nd] [F6] **Clean Up** and select **2:NewProb** to clear variables and set other defaults. In the MODE dialog box, set **Graph = FUNCTION**.
2. Enter the commands **Define f(x)=e^x** and **Define p(x)=a*x^2+b*x+c** to define the functions.

   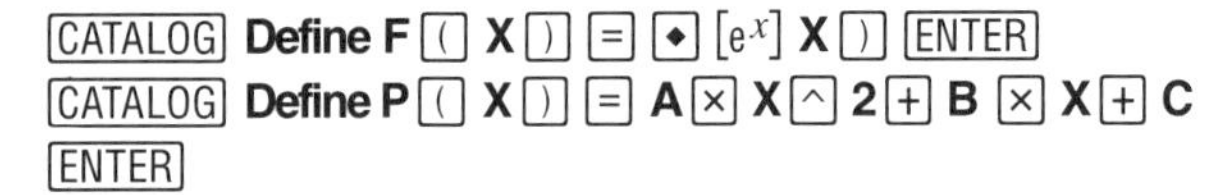

3. Enter **solve(f(x)=p(x), c)1x=0** to find the value of c.

   [CATALOG] **solve(F** [(] **X** [)] [=] **P** [(] **X** [)] [,] **C** [)] [|] **X** [=] **0**
   [ENTER]

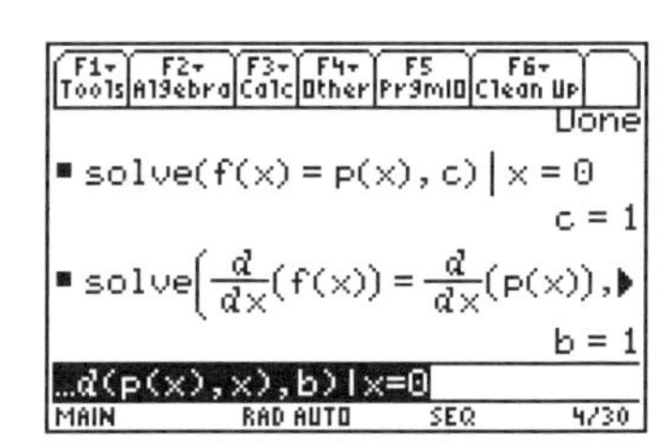

4. Solve for the coefficient b that will make $f'(0) = p'(0)$ with the command **solve(*d*(f(x),x)= *d*(p(x),x),b)|x=0**.

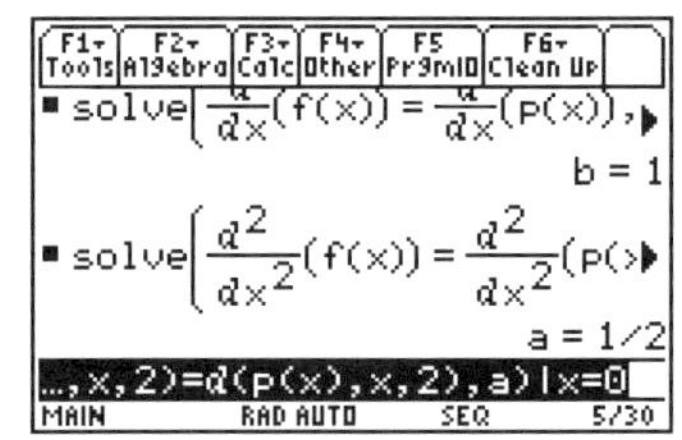

5. Solve for the coefficient a that will make $f''(0) = p''(0)$ with the command **solve(*d*(f(x),x,2)= *d*(p(x),x,2),a)|x=0**.

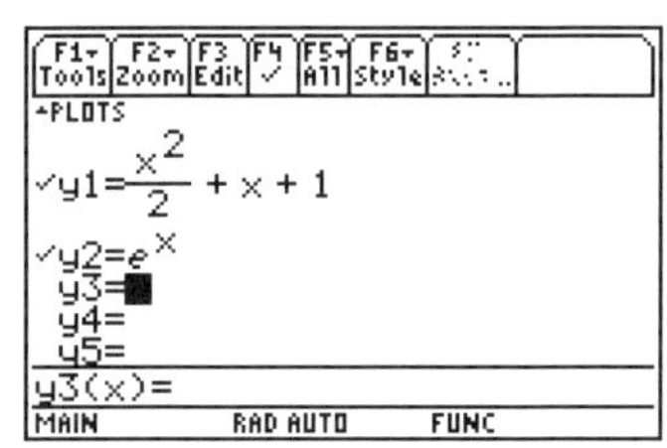

6. The coefficients are: $c = 1$, $b = 1$, $a = ½$. In the Y = Editor, enter the corresponding polynomial in $y1$. Enter e^(x) in $y2$.
7. Set up a [-4,4] x [-2,10] viewing window.
8. Press [◆] [GRAPH] to compare the graphs of the two functions.

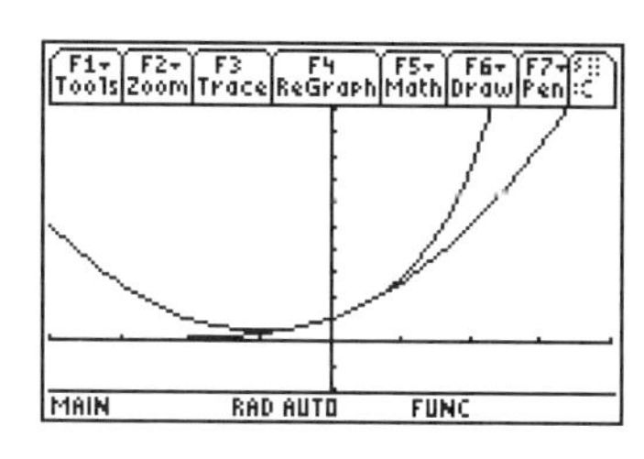

Notice that the polynominal is a good approximation of the exponential function for values of x near 0. This polynominal is called a 2nd *degree Taylor polynomial.*

You can generate this polynomial with the **taylor(** command from the **Calc** menu or the CATALOG. The parameters for this command are the function being approximated, the independent variable, the order of the polynomial, and the x-coordinate at which the polynomial is tangent to the function. If the last parameter is omitted, the TI-89 default value is 0.

9. Return to the Home screen and generate this polynomial with the **taylor(** command.

   [CATALOG] **taylor(** [◆] [e^x] **X**[)] [,] **X**[,] **2**[,] **0**[)] [ENTER]

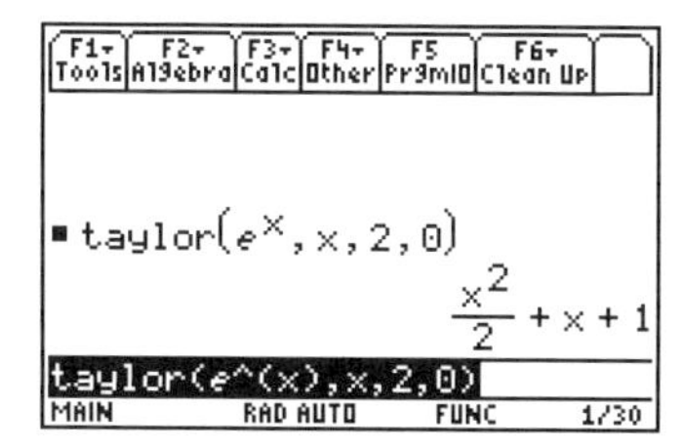

Example 4: Integral of a Taylor polynomial

Find the 9th degree Taylor polynomial for

$$f(x) = \tan^{-1} x$$

and compare with the integral of the 8th degree Taylor polynomial for

$$f(x) = \frac{1}{1+x^2}$$

Solution

1. Press [2nd] [F6] **Clean Up** and select **2:NewProb** to clear variables and set other defaults.

2. Enter the command **taylor(tan-1(x),x,9)**.

   [CATALOG] **taylor(** [◆] [TAN^{-1}] **X**[)] [,] **X**[,] **9**[ENTER]

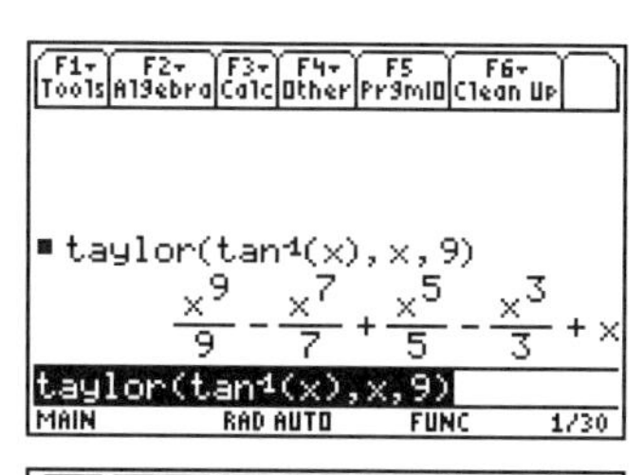

3. Now enter the command **∫(taylor(1/(1+x^2),x,8),x)**.

   [2nd] [∫] [CATALOG] **taylor(1** [÷] [(] **1** [+] **X**[^] **2**[)] [)] [,] **X**[,] **8**[)] [,] **X**[)] [ENTER]

   Both commands produce the same result. This was expected since

   $$\int \frac{1}{1+x^2} dx = \tan^{-1} x + C.$$

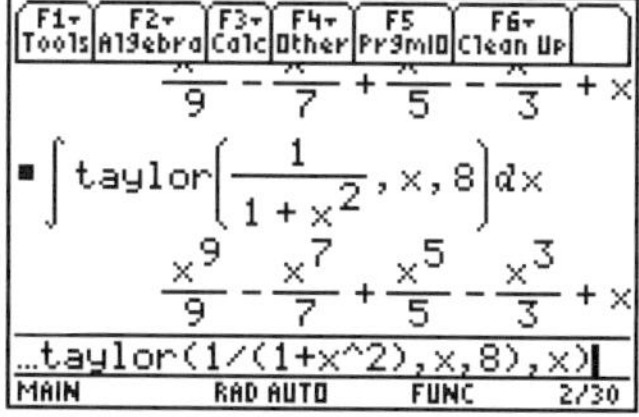

Exercises

Determine whether the sequences in exercises 1 through 3 converge or diverge. If the sequence converges, estimate the limit to which it converges. Support your conclusion graphically, numerically, and symbolically.

1. $a_n = \dfrac{(-1)^{n-1}}{n}$

2. $a_n = \dfrac{n!}{n^n}$

3. $a_n = \dfrac{4n^2 - 1}{3n + 2}$

Determine whether the series in exercises 4 through 8 converge or diverge. If a series converges, estimate the sum. Support your conclusion graphically, numerically, and symbolically.

4. $\sum_{k=1}^{\infty} \dfrac{(-1)^k}{k}$

5. $\sum_{k=0}^{\infty} \dfrac{1}{k!}$

6. $\sum_{k=0}^{\infty} 4\dfrac{(-1)^k}{2k+1}$

7. $\sum_{k=1}^{\infty} \left(\dfrac{2}{3}\right)^k$

8. $\sum_{k=1}^{\infty} \left(\dfrac{4}{3}\right)^{k-1}$

9. Find the 8th degree Taylor polynomial for $f(x) = \dfrac{1 + \cos 2x}{2}$ expanded about $x = 0$.

10. Find the 8th degree Taylor polynomial for $f(x) = \cos(2x)$.

    Add 1 to this result and then divide by 2. Use the **expand(** command to expand this result and compare it with the answer to Exercise 9.

11. Find the 5th degree Taylor polynomial for $f(x) = \ln(x)$ expanded about $x = 1$.

12. Find the 4th degree Taylor polynomial for $f(x) = \sin(x)$ expanded about $x = \dfrac{\pi}{2}$.

Appendix A

TI-89 Keystrokes and Menus

Using TI-89 keys

The TI-89 primary keys are mainly used for numbers, arithmetic operations, commonly-used variables such as x, and controls like the [ENTER] key. Primary keys are indicated in this book with a key font, with the exception of **X, Y, Z, T** and numbers, which are displayed as text.

[2nd] functions

Most keys have a second function, indicated above the key in yellow. To access these functions, press the yellow [2nd] key, and then the primary key. For example, to enter **cos**, press [2nd] and then the **Z** key, because **COS** is in yellow above **Z**. These keystrokes will be indicated in this book as [2nd] [COS] with the brackets indicating that the command is the second and not the primary function of the key.

[◆] functions

Some keys also have another function in green. To access these functions, press the green [◆], and then the primary key. For example, to enter **cos**$^{-1}$, press the green [◆] and then the **Z** key because **COS**$^{-1}$ is in green above **Z**. These keystrokes will be indicated in this book as [◆] [COS^{-1}], with the brackets indicating that the command is the green function and not the primary function of the key. There are additional green functions that do not appear on the keyboard. Press [◆] [K] to see these choices. For example, you will see that [◆] [÷] is **!**.

[alpha] functions

Most keys that do not have a green function have a purple letter above them. To access these letters, press the purple [alpha] key and then the primary key. For example, to enter **A**, press the [alpha] key and then the [=] key because **A** is in purple above [=]. These keystrokes will be indicated in this book as **A**. If you are typing several letters in a row, you can use [2nd] [a-lock], which will eliminate the need to press [alpha] for each letter. Press [alpha] at any time to release alpha-lock. The letters **X, Y ,Z, T** are primary keys and do not need [alpha]. If you forget and press [alpha], those keys will still enter the appropriate character. That is, [alpha] **T** is the same as [T]. The [↑] is a shift key to type upper-case letters. Press [↑] and the key for a single upper-case letter. If you press [↑] [alpha] you will be in upper-case alpha-lock. The [␣] is a space.

Function keys / Toolbar menus

The keys [F1] through [F5] in the top row access different menus or features indicated across the top of the screen. These choices vary depending on the Editor you are using. For instance, if you are on the Home screen, [F3] is the **Calculus** menu; however, if you are on the Graph screen, then [F3] is **Trace**. In some cases, there will be additional choices for [F6], [F7], or [F8]. These choices require two keystrokes, for example, [F6] is the second function of [F1]. In this book, these keys will be shown by key and by name such as [F1] **Tools**, or [2nd] [F6] **Clean Up**.

In many cases pressing the function keys will display a toolbar menu. The choices can be accessed by typing the number or letter in front of the command or by highlighting the choice and pressing [ENTER]. For example, pressing **3** will get a **limit(** command from the menu shown here. In this book, this process will be shown as: Press [F3] **Calc** and select **3:limit(**. The down arrow symbol after item 8 indicates that there are additional commands not currently visible on the screen.

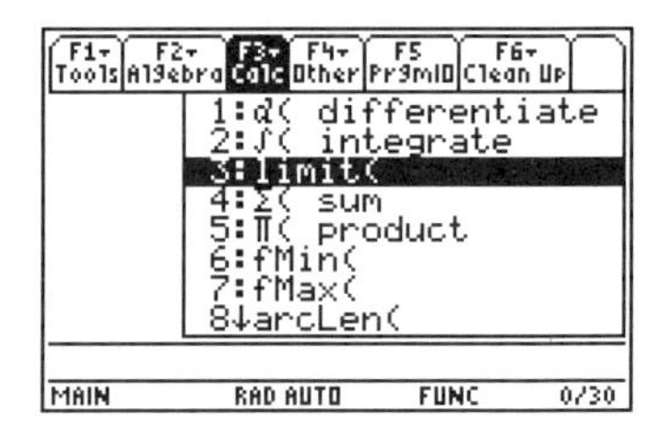

In some cases, a menu item may have an arrowhead at the right of the item. See the **MATH** menu pictured here. In this case, press ▶ to display a submenu.

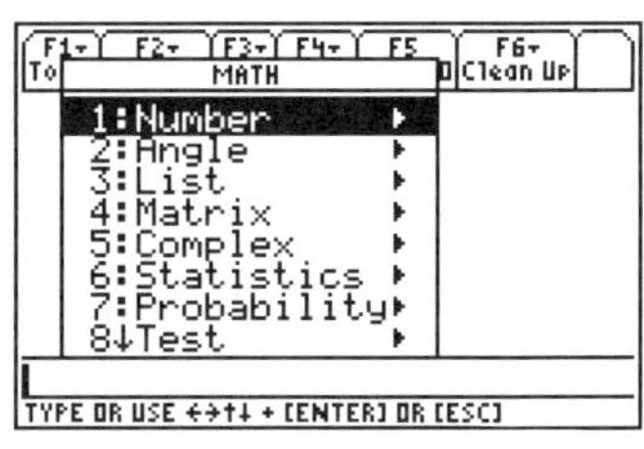

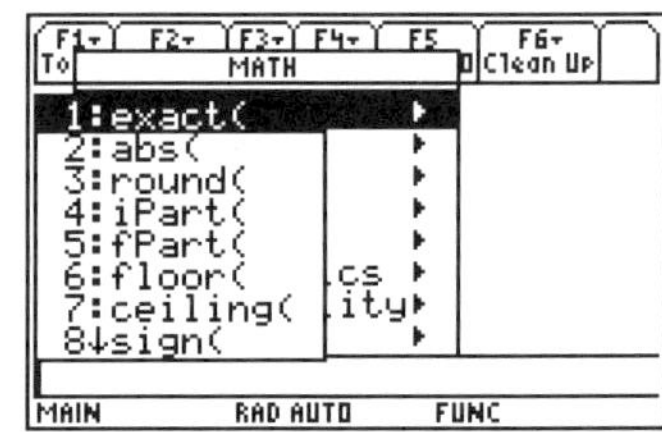

CATALOG

Most of the TI-89 commands are also found in the CATALOG. Press [CATALOG], and then press the key corresponding to the first letter of the command. For example, press **D** to move to commands that begin with *d*. (Do not press [alpha].) Press [2nd] ⊖ to "page down" through the CATALOG. Notice that on-screen help is provided in the CATALOG. The appropriate parameters for that function are shown in the bottom line of the screen (the "status bar"). Parameters in [ ] are optional. This on-screen help is not shown in other menus. To use a command fron the CATALOG, move the ⊙ indicator to the command and press [ENTER].

Basic editing

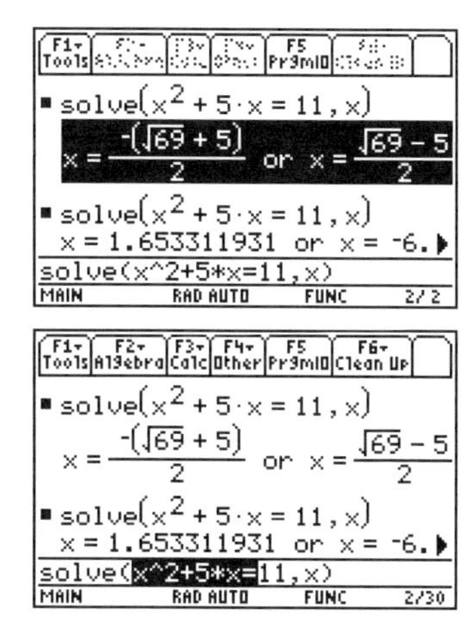

On the Home screen, you can use ⊖ and ⊖ to travel through the history area and highlight any command or result. Press [ENTER] to bring any command or result back to the entry line.

Press ⊙ or ⊙ to move the cursor when you are on the edit line. Press [2nd] ⊙ to move to the beginning of the line, and press [2nd] ⊙ to move to the end. The thin cursor is an insert cursor. Press [2nd] [INS] to change to a larger, overstrike cursor. Press [2nd] [INS] to toggle back and forth between these choices. Press [◆] [DEL] or [←] to delete the character at the left of the cursor.

Press and hold [↑] and use ⊙ or ⊙ to highlight any part of the entry line. Use [F1] **Tools** for Copy and Paste commands.

Appendix B

Common Calculus Operations

Before you enter the following examples, you should reset your TI-89 to its default settings by pressing [2nd] [MEM] [F1] **3:Default** [ENTER] [ENTER] and then clear all one-letter variables by pressing [2nd] [F6] **1:Clear a-z** [ENTER].

If you want to clear the Home screen and the entry line before beginning a new example, press [HOME] to move to the Home screen and then press [F1] **8:Clear Home** [CLEAR].

Graphing functions

Graph $y = x^2 - 2$ in a [-5,5] x [-5,10] window.

1. Enter the function in the Y= Editor.

   [♦] [Y=] [CLEAR] **X** [^] **2** [−] **2** [ENTER]

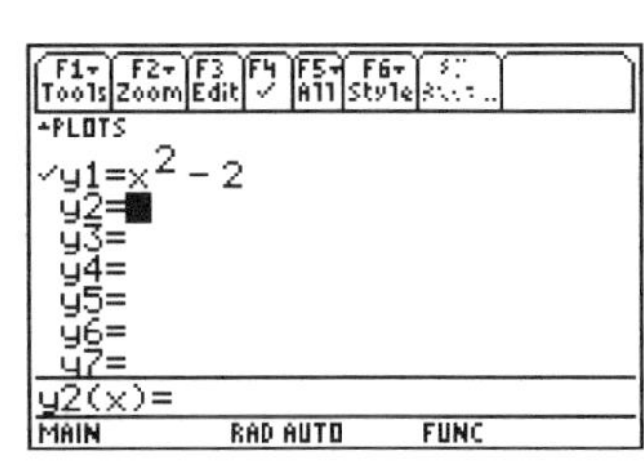

2. Select the Window Editor by pressing [♦] [WINDOW].
3. Enter the x-window values:

   [(-)] **5** [ENTER] **5** [ENTER] **1** [ENTER]

4. Enter the y-window values:

   [(-)] **5** [ENTER] **10** [ENTER] **1** [ENTER]

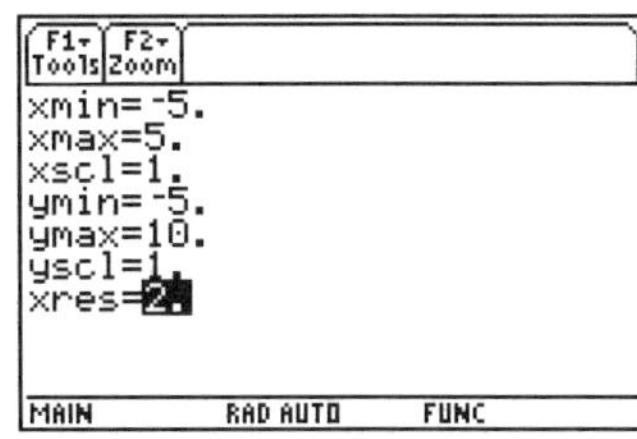

5. Graph the function by pressing [♦] [GRAPH].

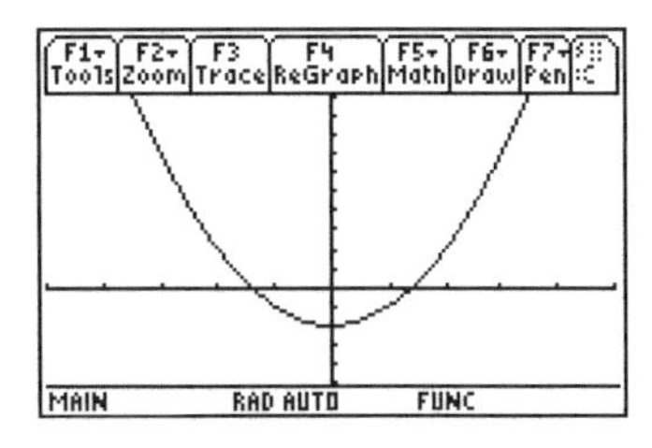

Limits

Find $\lim_{x \to \infty} (e^{-x})$.

From the Home screen, press:

[CATALOG] **limit(** [◆] [e^x] [(-)] **X** [)] [,] **X** [,] [◆] [∞] [)] [ENTER]

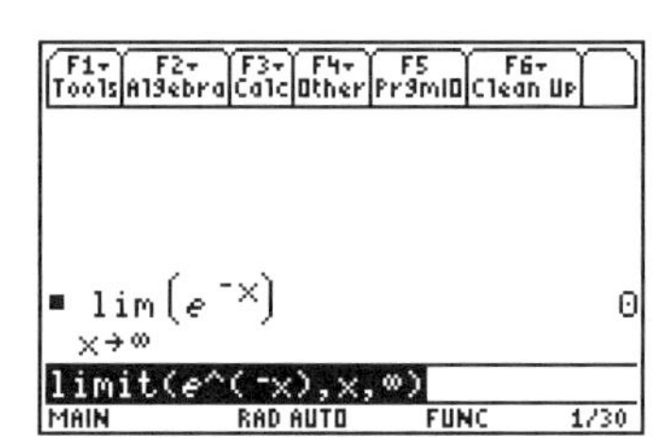

The first derivative of a function

Take the first derivative of $y = \frac{1}{x}$ with respect to x.

From the Home screen, press:

[HOME] [2nd] [d] **1** [÷] **X** [,] **X** [)] [ENTER]

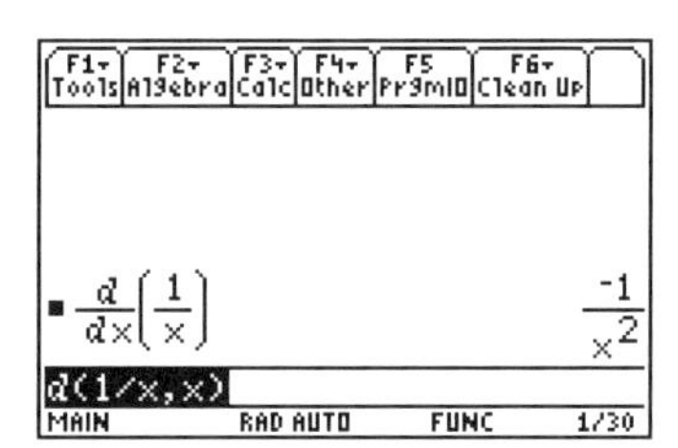

The second derivative of a function

Take the second derivative of $y = \frac{1}{x}$ with respect to x.

From the Home screen, press:

[HOME] [2nd] [d] **1** [÷] **X** [,] **X** [,] **2** [)] [ENTER]

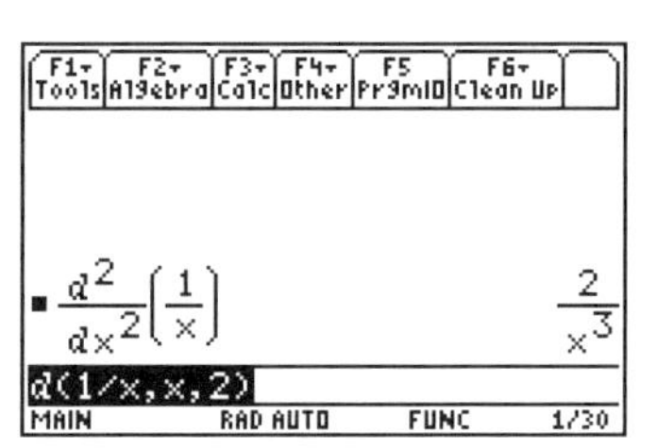

The roots of an equation

Find the real roots of $y = x^2 + 3x - 5$.

From the Home screen, press:

[CATALOG] **zeros(** **X** [^] **2** [+] **3** **X** [-] **5** [,] **X** [)] [ENTER]

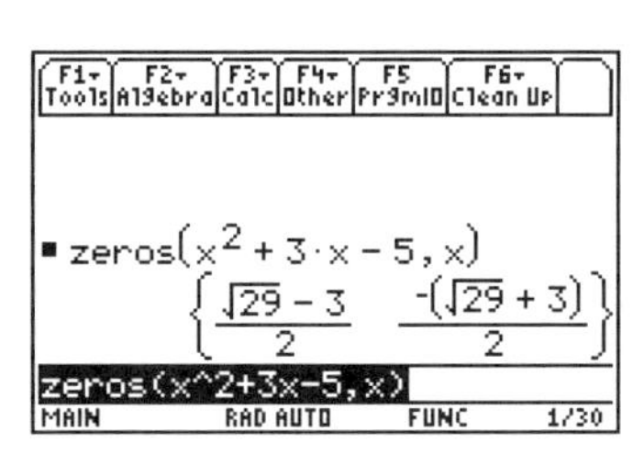

Indefinite integrals

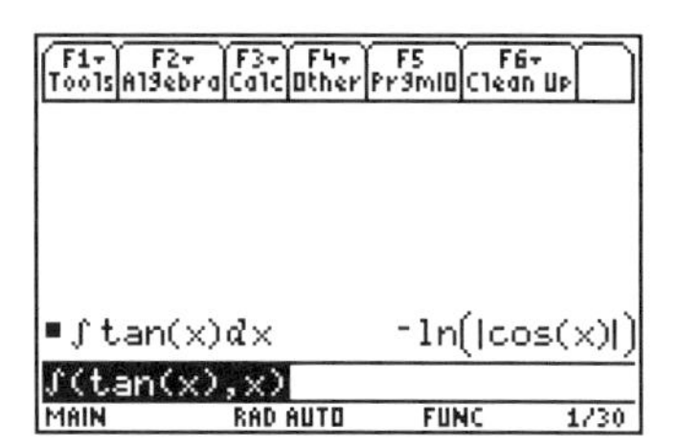

Evaluate $\int \tan(x)dx$.

From the Home screen, press:

[2nd] [∫] [2nd] [TAN] **X** [)] [,] **X** [)] [ENTER]

Definite integrals

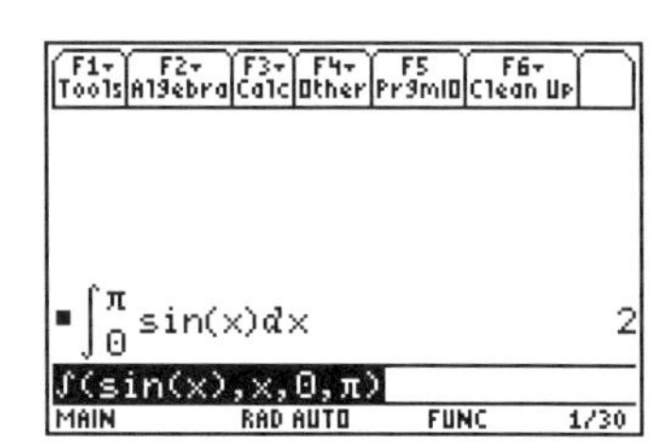

Evaluate $\int_0^{\pi} \sin x dx$.

From the Home screen, press:

[2nd] [∫] [2nd] [SIN] **X** [)] [,] **X** [,] **0** [,] [2nd] [π] [)] [ENTER]

Series

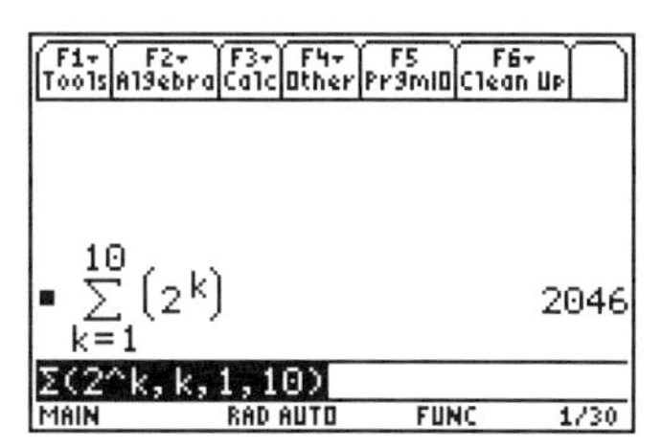

Evaluate $\sum_{k=1}^{10} 2^k$.

From the Home screen, press:

[CATALOG] **Σ(2** [^] [alpha] **K** [,] [alpha] **K** [,] **1** [,] **10** [)] [ENTER]

Taylor series

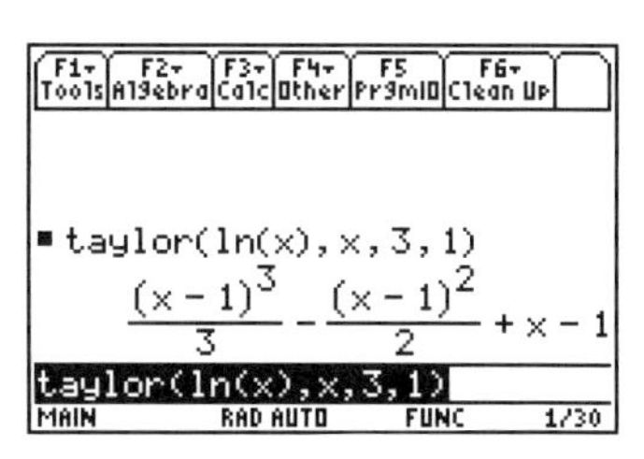

Find the third degree Taylor polynomial for $y = \ln x$ expanded about $x = 1$.

From the Home screen, press:

[CATALOG] **taylor(** [2nd] [LN] **X** [)] [,] **X** [,] **3** [,] **1** [)] [ENTER]

Appendix C

Creating Scripts

Scripts provide a way to save a sequence of commands that have been entered on the Home screen. This sequence then can be presented at a later time by executing the script. This can be a useful way to prepare a lesson or the solution to a problem in advance and then present it quickly with few key strokes.

Example: Inductive chain rule lesson

Students can discover the chain rule by seeing many examples. These examples can be prepared and entered as a script before class and then be presented quickly in class.

Here are the steps to create the script:

1. Press [2nd] [F6] **Clean Up** and select **2:NewProb** to clear variables and set other defaults. Press [F1] **Tools** and select **8:Clear Home** to clear the Home screen. Press [ENTER] to clear the entry line.
2. Enter the following derivatives on the Home screen:

   d(sin(x),x)

   d(sin(2x),x)

   d(sin(3x),x)

   d(sin(4x),x)

   d(sin(x^2),x)

   d(sin(x^3),x)

   d(sin(x^2+2x),x)

   d(sin(cos(x)),x)

   d(sin(f(x)),x)

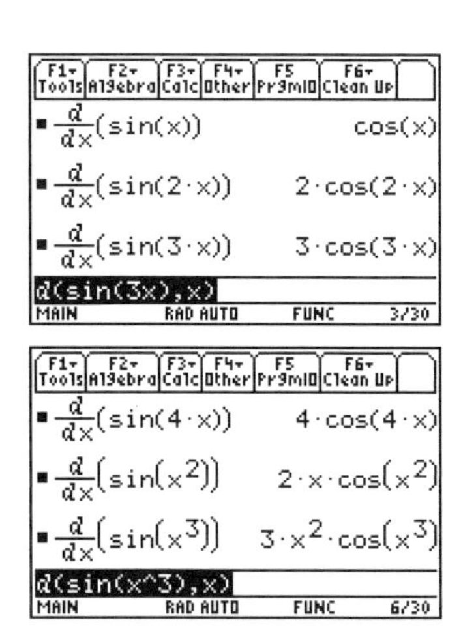

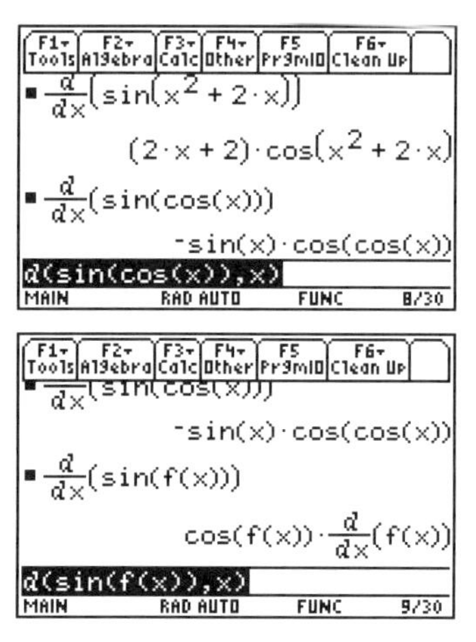

3. Press [F1] **Tools** and select **2:Save Copy As.**

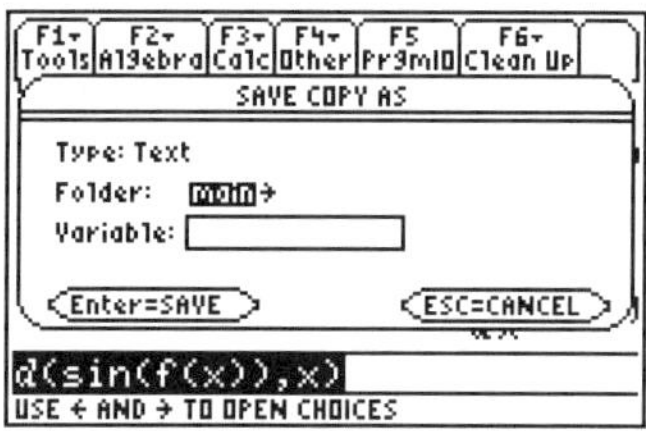

4. Press ⊙ to move the cursor to the box labeled **Variable** and enter the name **chain** for this script. Put the calculator in alpha-lock mode by pressing [2nd] [a-lock] before entering the letters of the name.

5. Press [ENTER] twice to save the name and the script and press [alpha] to turn off alpha-lock. Then press [F1] **Tools** and select **8:Clear Home** to clear the Home screen. Then press [CLEAR] to clear the entry line.

Here are the steps to present this script:

1. Press [APPS] and select **8:Text Editor.** Then select **2:Open.**

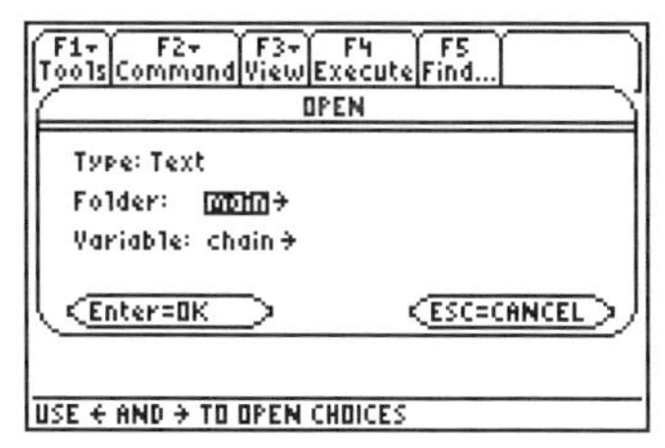

2. You can select the script to be executed from this dialog box. Since **chain** is already selected, press [ENTER].

3. Split the screen by pressing [F3] **View** and selecting **1:Scriptview**.

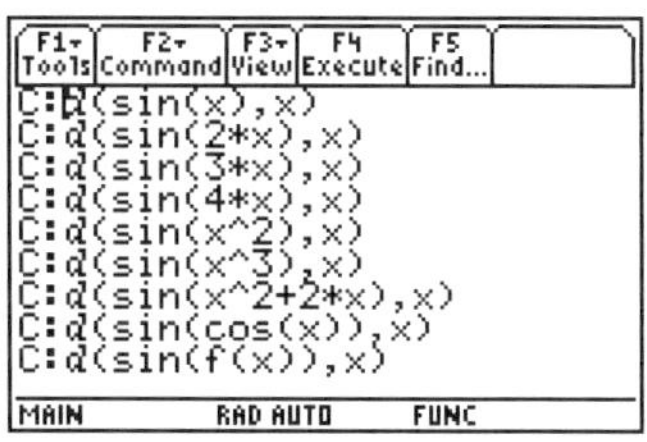

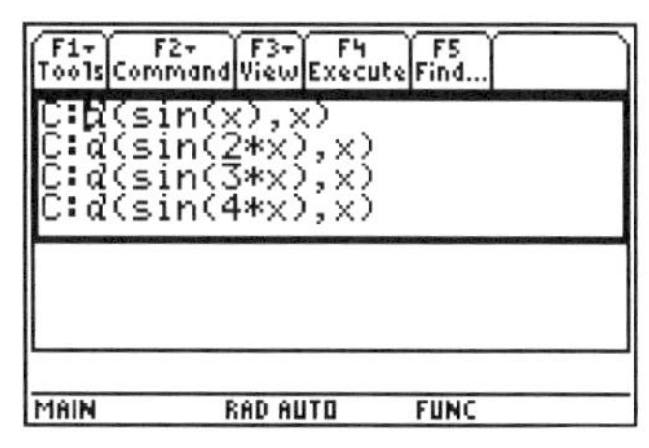

4. Execute the first command by pressing [F4] **Execute**. The result of this command appears in the bottom half of the screen.

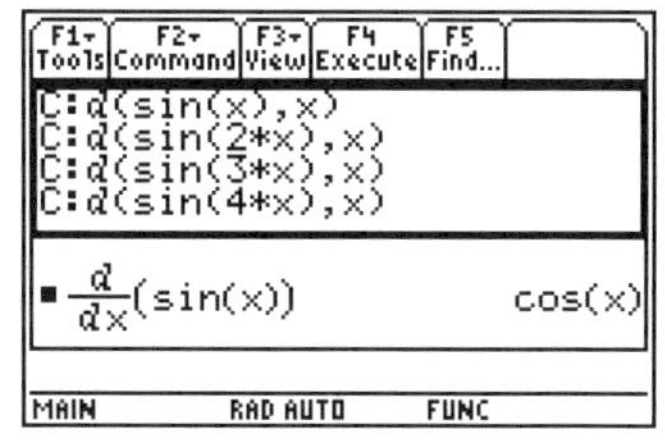

5. The cursor on the top half of the screen should have moved to the next line. Each time you press [F4], another command is executed and the result is displayed on the bottom half of the screen. You should ask students for their predictions each time before you execute a script command. After several commands have been executed, ask the students to write the patterns they see. You can continue in this fashion until all the commands have been executed.

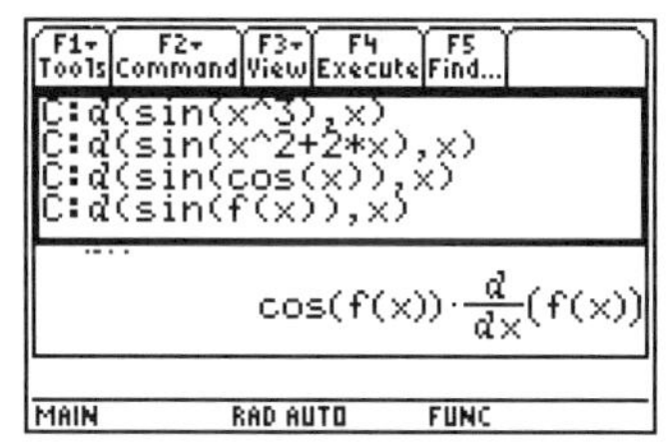

6. Ask students to write a brief summary of the rule they have discovered. Have them try it on other functions besides $y = \sin(x)$. You can clear the split screen and return to the Home screen by pressing [F3] **View**, selecting **2:Clear split**, and pressing [HOME].

   Be careful while you are in the Text Editor executing a script, since any changes you make in this screen cause the script to be updated automatically.

   Teachers and students can use a technique similar to the one we have just described to create and present scripts that demonstrate solutions to complicated problems.

Appendix D

Solutions to the Exercises

Chapter 1

1. 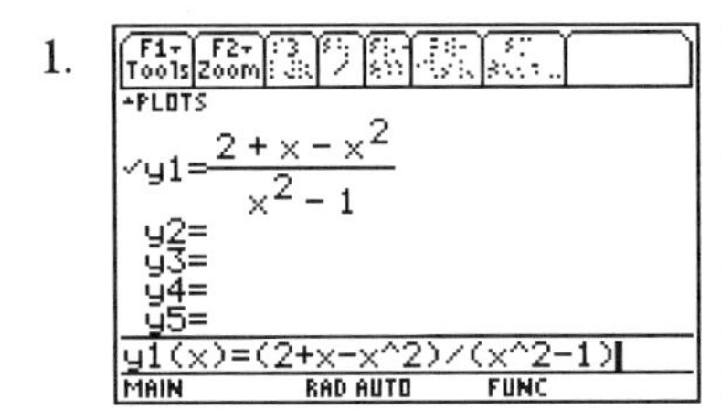

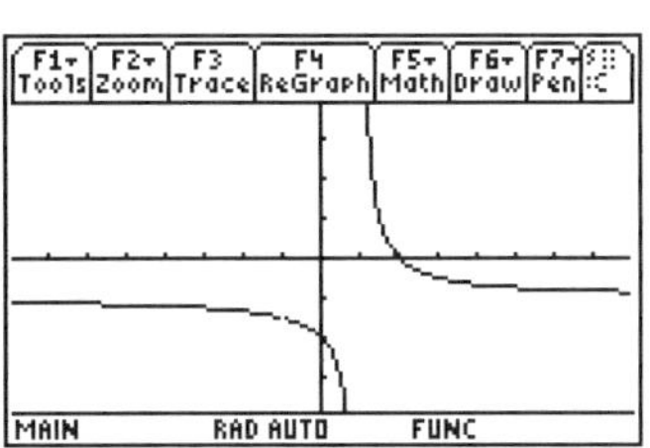

2. 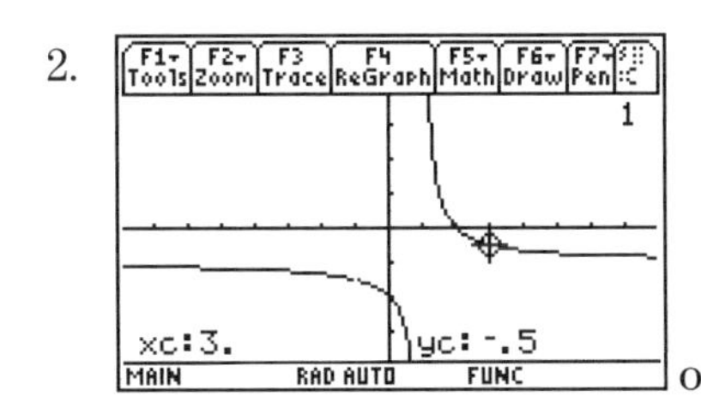

or
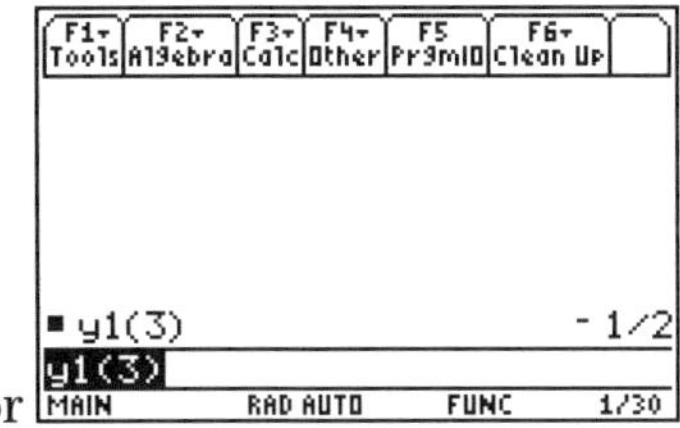

3. 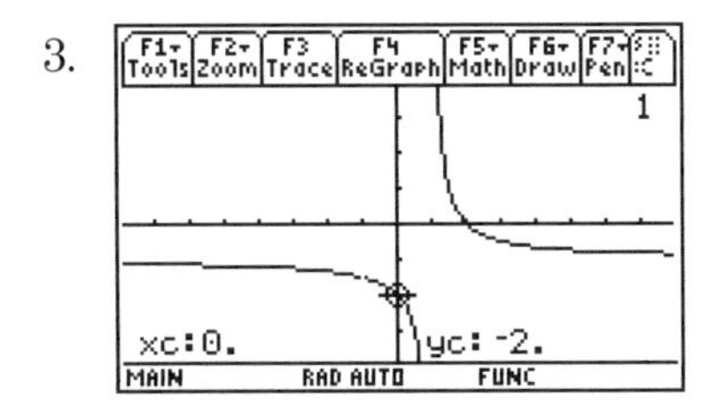

4.

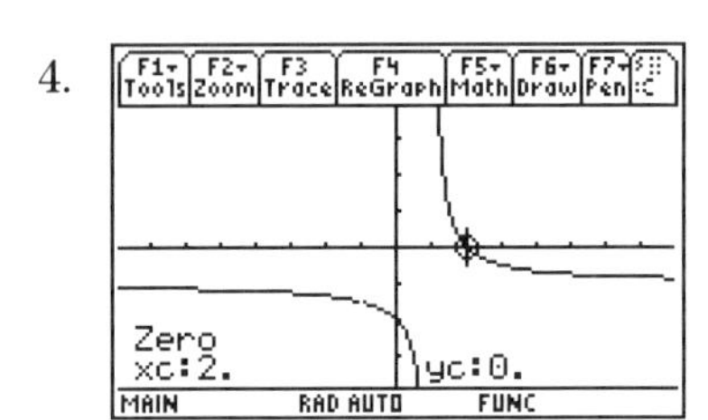

5.

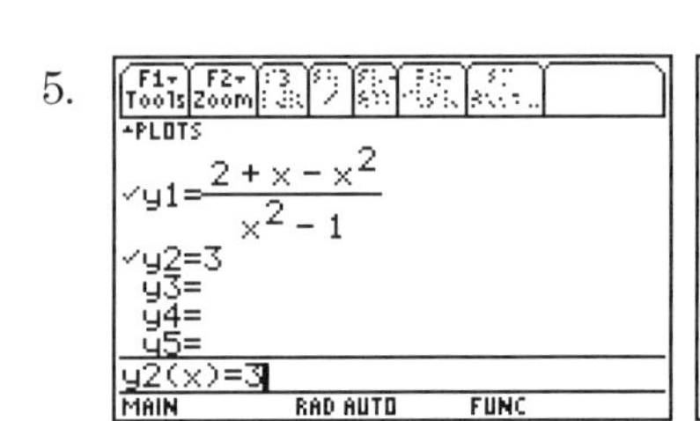

Intersection
xc:1.25
yc:3.
MAIN RAD AUTO FUNC

6.

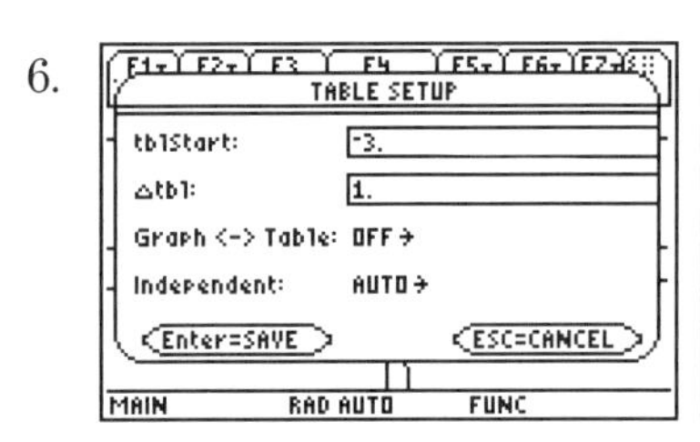

x	y1
-3.	-1.25
-2.	-1.333
-1.	undef
0.	-2.
1.	undef

x=-3.

x	y1
-1.	undef
0.	-2.
1.	undef
2.	0.
3.	-.5

x=3.

7.

$\lim_{x \to -1} y1(x)$ -3/2

limit(y1(x),x,-1)

MAIN RAD AUTO FUNC 1/30

8.

9.

$\lim_{x \to \infty} y1(x)$ -1

$\lim_{x \to -\infty} y1(x)$ -1

limit(y1(x),x,-∞)

MAIN RAD AUTO FUNC 2/30

Chapter 2

Results are shown with **Display Digits = FLOAT 10**.

1. 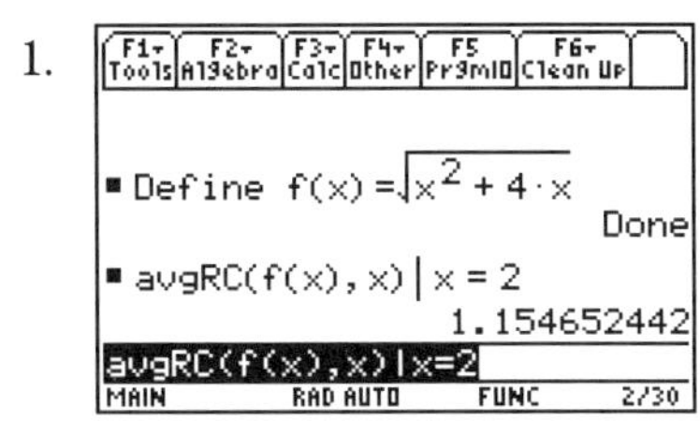

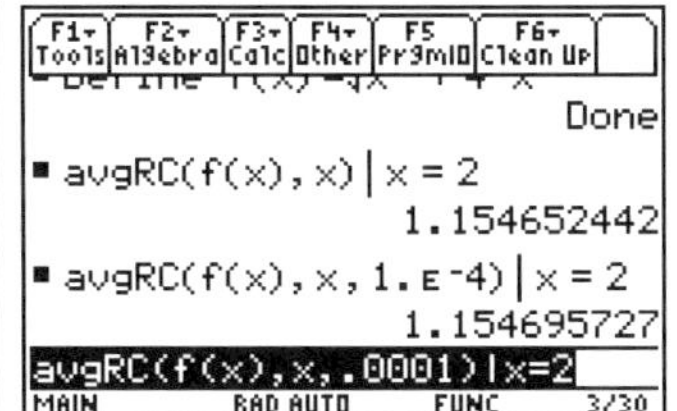

2. 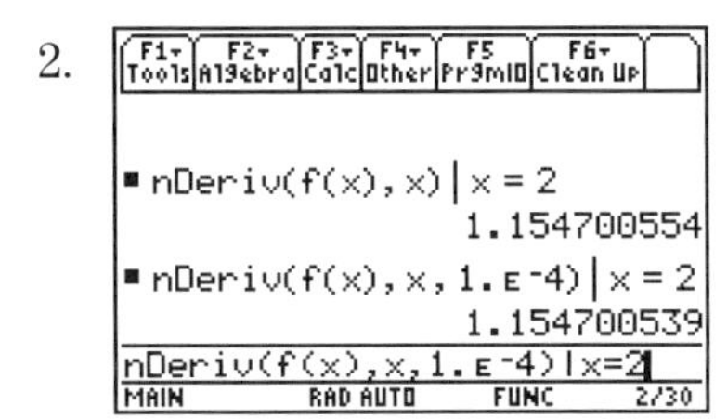

3. 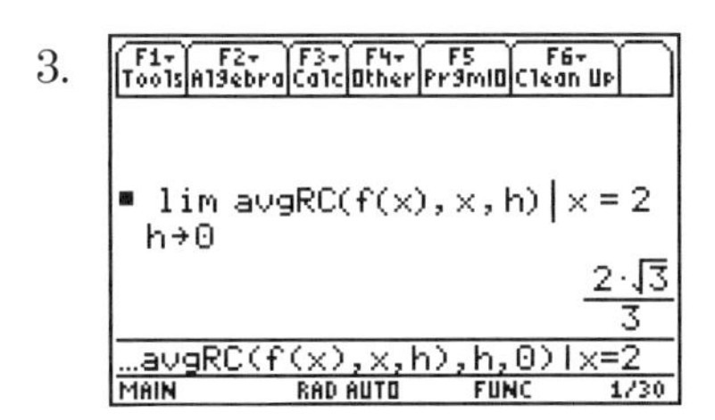

4. 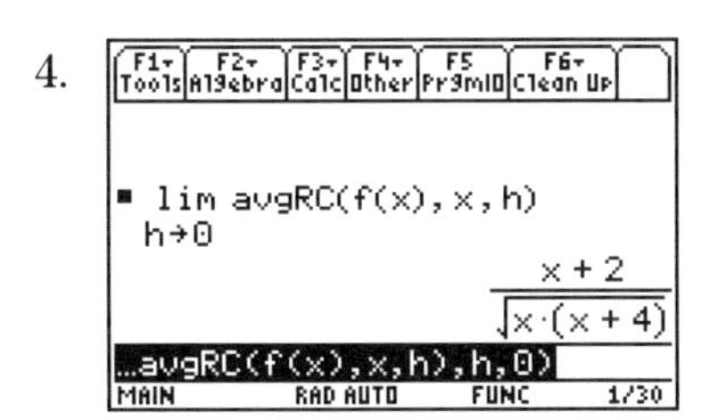

5. 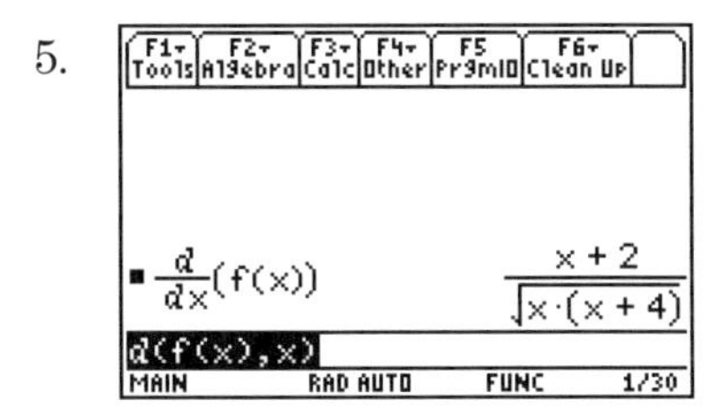

6.

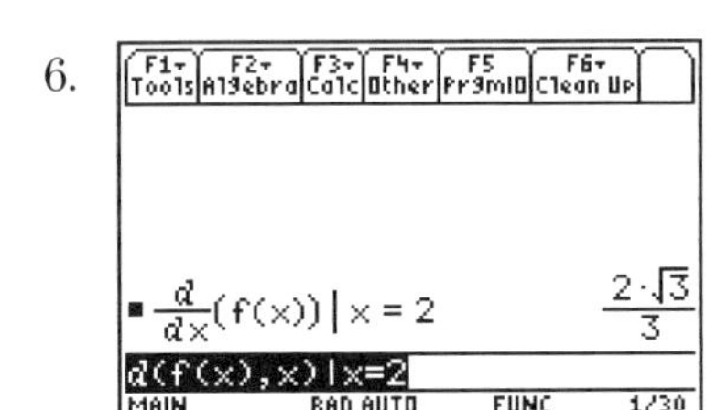

7.

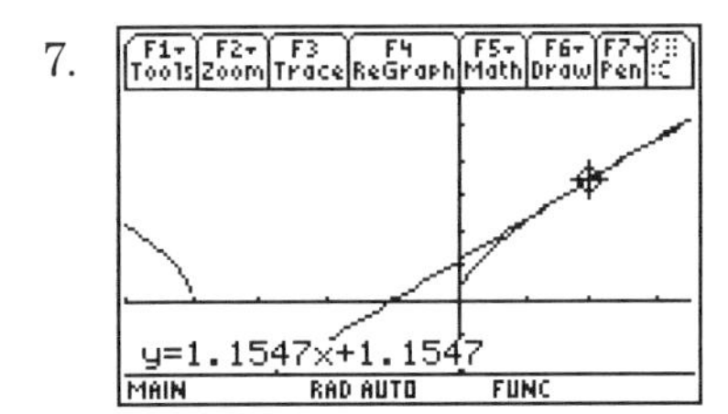

8.

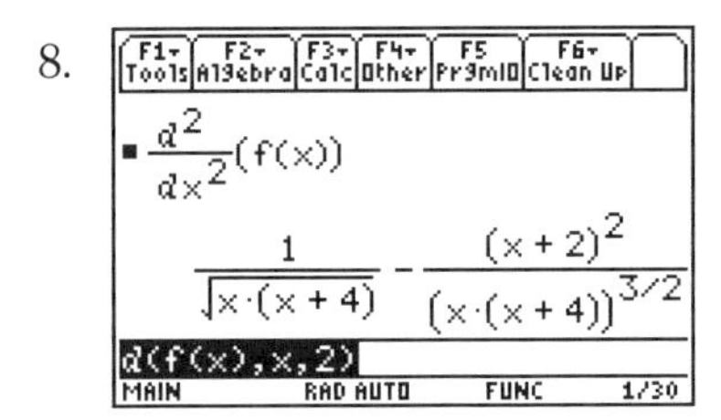

9.

F1 Tools F2 Algebra F3 Calc F4 Other F5 PrgmIO F6 Clean Up

■ $3\cdot x^2 - x\cdot y + y^2 - 4\cdot x + 5\cdot y$ ▸

$3\cdot x^2 + x\cdot(-y - 4) + y^2 + 5\cdot y$ ▸

3x^2-x*y+y^2-4x+5y-10→z

MAIN RAD AUTO FUNC 1/30

F1 Tools F2 Algebra F3 Calc F4 Other F5 PrgmIO F6 Clean Up

■ $\dfrac{-\frac{d}{dx}(z)}{\frac{d}{dy}(z)}$ $\quad \dfrac{y - 2\cdot(3\cdot x - 2)}{2\cdot y - x + 5}$

-d(z,x)/d(z,y)

MAIN RAD AUTO FUNC 1/30

10.

F1 Tools F2 Algebra F3 Calc F4 Other F5 PrgmIO F6 Clean Up

■ solve(z = 0, y) | x = -2

y = -2 or y = -5

solve(z=0,y)|x=-2

MAIN RAD AUTO FUNC 1/30

11.

F1 Tools F2 Algebra F3 Calc F4 Other F5 PrgmIO F6 Clean Up

■ solve(z = 0, y)

$y = \dfrac{-(\sqrt{-11\cdot x^2 + 6\cdot x + 65} - x}{2}$ ▸

solve(z=0,y)

MAIN RAD AUTO FUNC 1/30

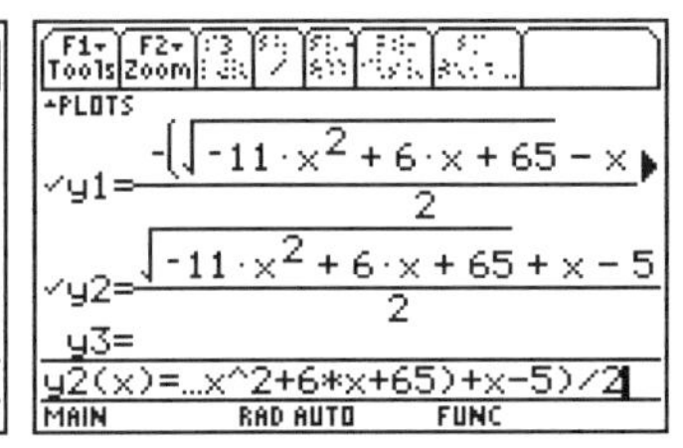

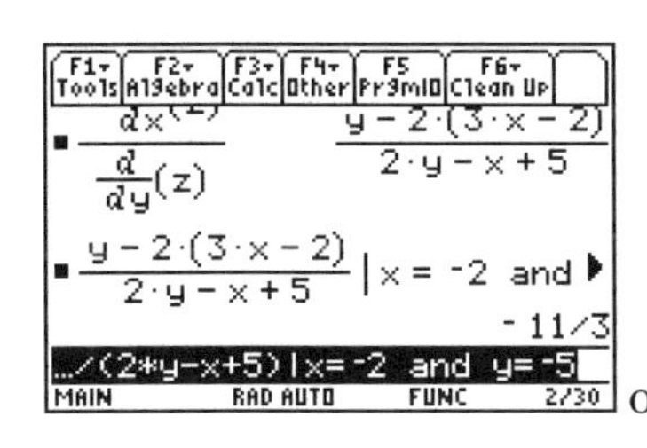
F1 Tools F2 Algebra F3 Calc F4 Other F5 PrgmIO F6 Clean Up
y − 2·(3·x − 2) / 2·y − x + 5
y − 2·(3·x − 2) / 2·y − x + 5 | x = -2 and
-11/3
…/(2*y-x+5)|x=-2 and y=-5
MAIN RAD AUTO FUNC 2/30
or
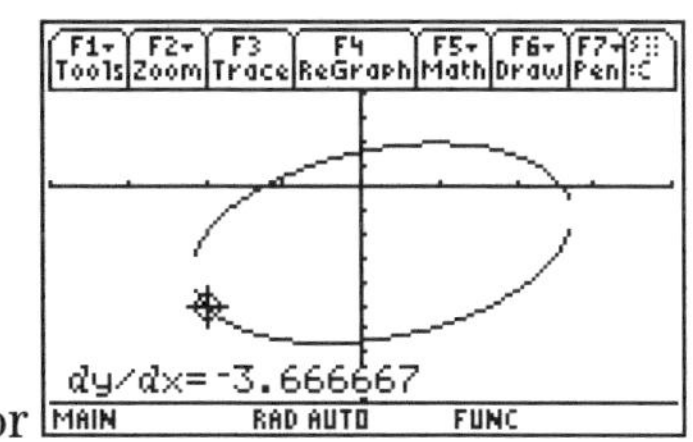
F1 Tools F2 Zoom F3 Trace F4 ReGraph F5 Math F6 Draw F7 Pen
dy/dx=-3.666667
MAIN RAD AUTO FUNC

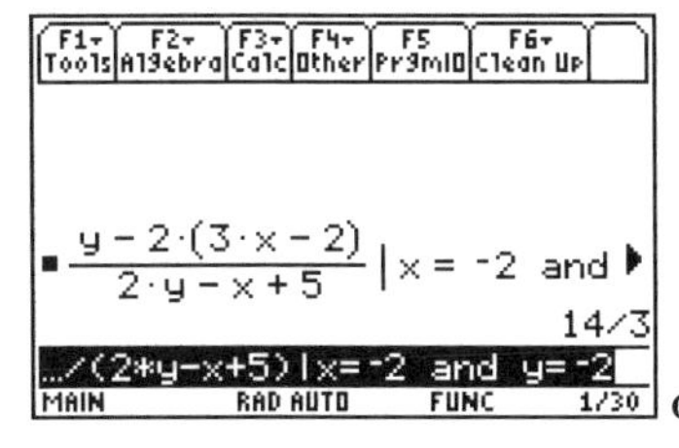
F1 Tools F2 Algebra F3 Calc F4 Other F5 PrgmIO F6 Clean Up
y − 2·(3·x − 2) / 2·y − x + 5 | x = -2 and
14/3
…/(2*y-x+5)|x=-2 and y=-2
MAIN RAD AUTO FUNC 1/30
or
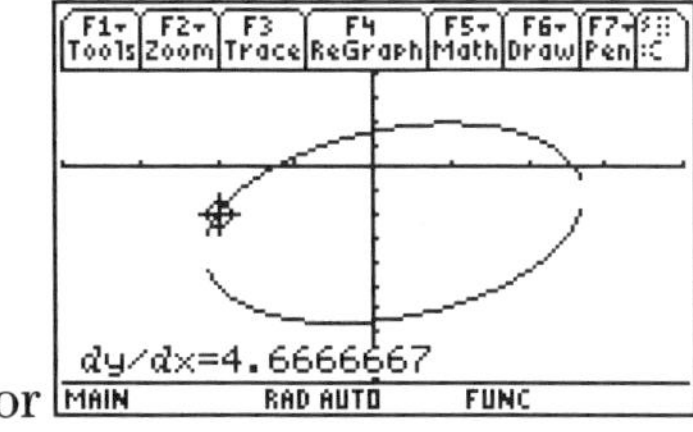
F1 Tools F2 Zoom F3 Trace F4 ReGraph F5 Math F6 Draw F7 Pen
dy/dx=4.6666667
MAIN RAD AUTO FUNC

12.

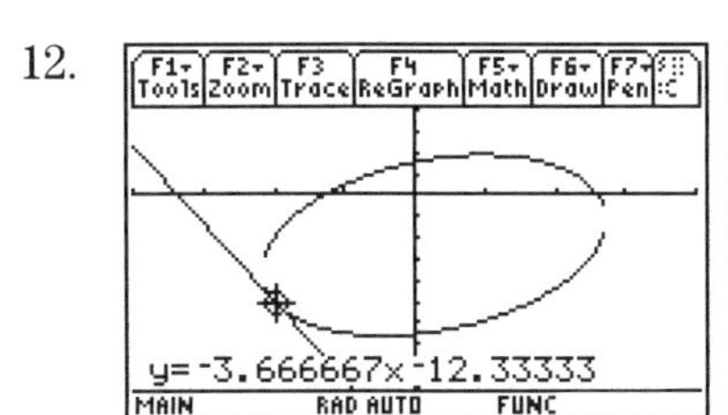
F1 Tools F2 Zoom F3 Trace F4 ReGraph F5 Math F6 Draw F7 Pen
y=-3.666667x-12.33333
MAIN RAD AUTO FUNC
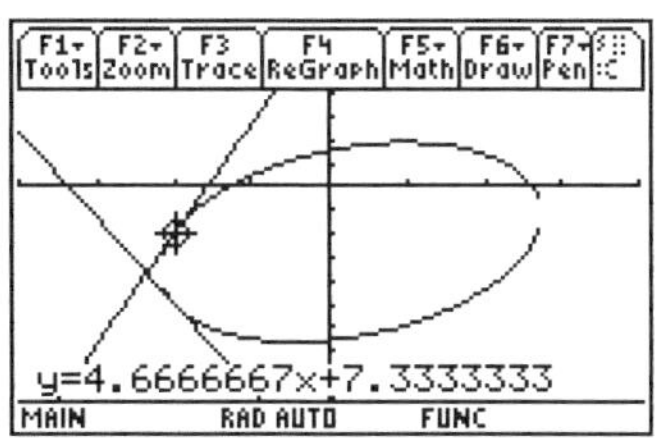
F1 Tools F2 Zoom F3 Trace F4 ReGraph F5 Math F6 Draw F7 Pen
y=4.6666667x+7.3333333
MAIN RAD AUTO FUNC

13.

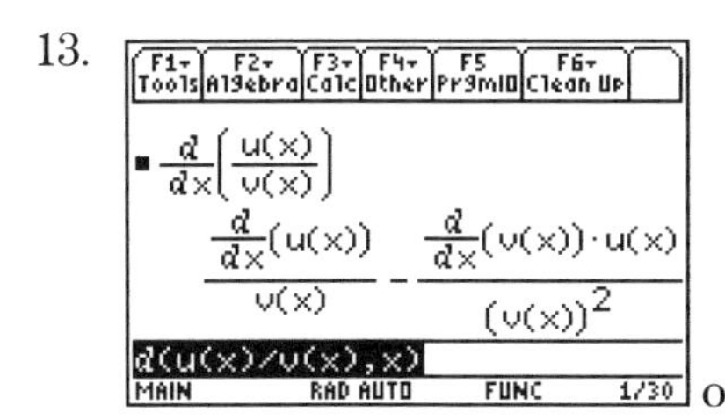
F1 Tools F2 Algebra F3 Calc F4 Other F5 PrgmIO F6 Clean Up
d(u(x)/v(x),x)
MAIN RAD AUTO FUNC 1/30
or
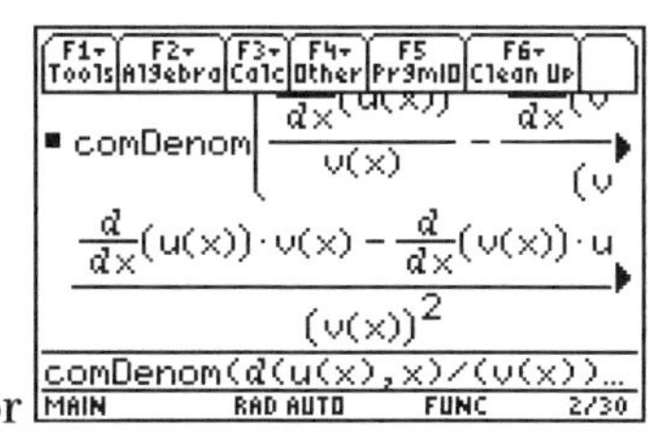
F1 Tools F2 Algebra F3 Calc F4 Other F5 PrgmIO F6 Clean Up
comDenom
comDenom(d(u(x),x)/(v(x))…
MAIN RAD AUTO FUNC 2/30

Chapter 3

1.

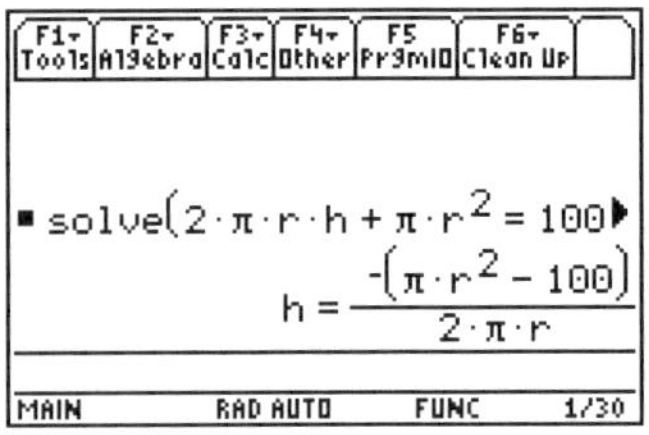

F1▾ Tools | F2▾ Algebra | F3▾ Calc | F4▾ Other | F5 PrgmIO | F6▾ Clean Up

■ $\text{solve}(2\cdot\pi\cdot r\cdot h + \pi\cdot r^2 = 100$ ▸ $\quad h = \dfrac{-(\pi\cdot r^2 - 100)}{2\cdot\pi\cdot r}$

MAIN RAD AUTO FUNC 1/30

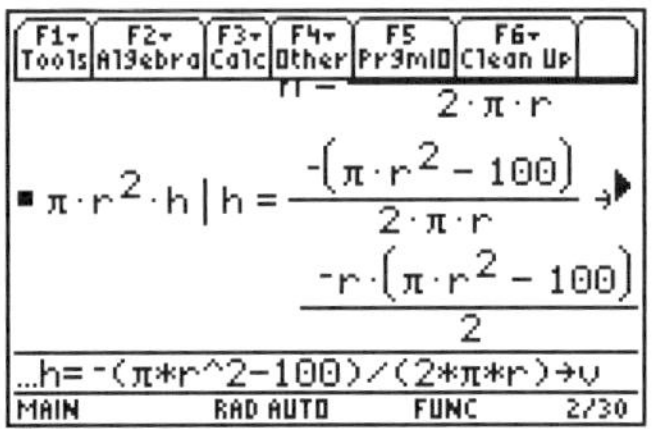

F1▾ Tools | F2▾ Algebra | F3▾ Calc | F4▾ Other | F5 PrgmIO | F6▾ Clean Up

$h = \dfrac{\quad}{2\cdot\pi\cdot r}$

■ $\pi\cdot r^2\cdot h \mid h = \dfrac{-(\pi\cdot r^2 - 100)}{2\cdot\pi\cdot r} \rightarrow$ ▸ $\quad \dfrac{-r\cdot(\pi\cdot r^2 - 100)}{2}$

...h=-(π*r^2-100)/(2*π*r)→v

MAIN RAD AUTO FUNC 2/30

2.

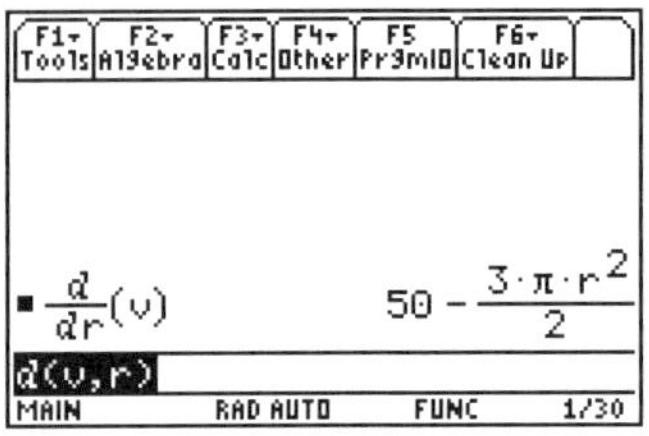

F1▾ Tools | F2▾ Algebra | F3▾ Calc | F4▾ Other | F5 PrgmIO | F6▾ Clean Up

■ $\dfrac{d}{dr}(v) \qquad 50 - \dfrac{3\cdot\pi\cdot r^2}{2}$

d(v,r)

MAIN RAD AUTO FUNC 1/30

3.

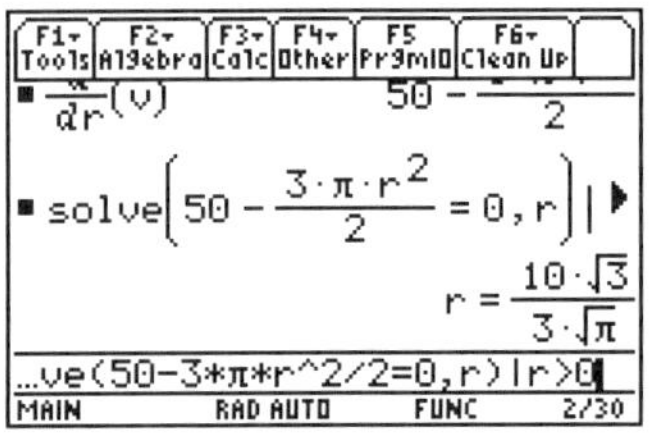

F1▾ Tools | F2▾ Algebra | F3▾ Calc | F4▾ Other | F5 PrgmIO | F6▾ Clean Up

■ $\dfrac{d}{dr}(v) \qquad 50 - \dfrac{\quad}{2}$

■ $\text{solve}\left(50 - \dfrac{3\cdot\pi\cdot r^2}{2} = 0, r\right) \mid$ ▸ $\quad r = \dfrac{10\cdot\sqrt{3}}{3\cdot\sqrt{\pi}}$

...ve(50-3*π*r^2/2=0,r)|r>0

MAIN RAD AUTO FUNC 2/30

4.

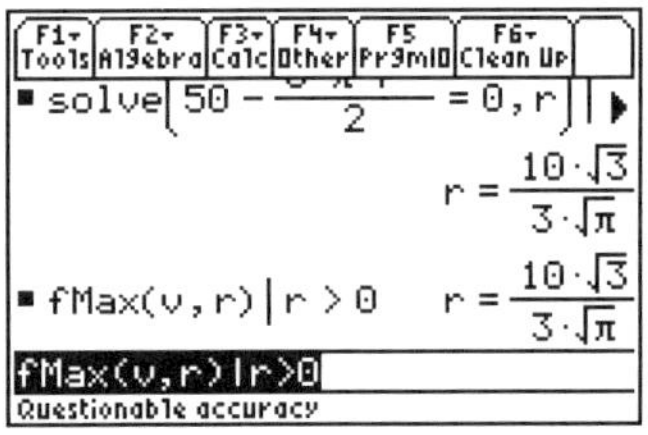

F1▾ Tools | F2▾ Algebra | F3▾ Calc | F4▾ Other | F5 PrgmIO | F6▾ Clean Up

■ $\text{solve}\left(50 - \dfrac{\quad}{2} = 0, r\right) \mid$ ▸ $\quad r = \dfrac{10\cdot\sqrt{3}}{3\cdot\sqrt{\pi}}$

■ $\text{fMax}(v, r) \mid r > 0 \qquad r = \dfrac{10\cdot\sqrt{3}}{3\cdot\sqrt{\pi}}$

fMax(v,r)|r>0

Questionable accuracy

5.

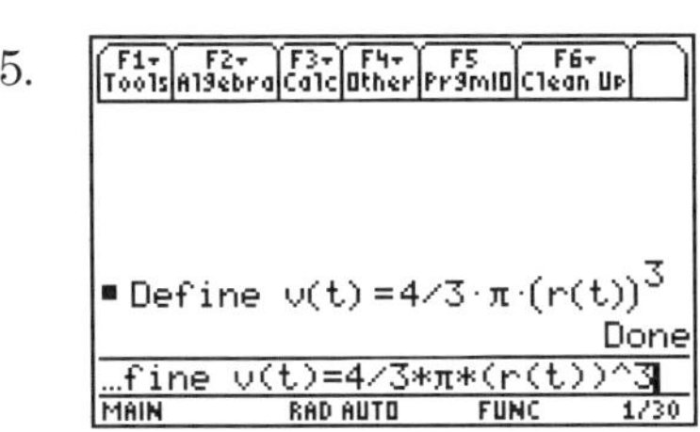

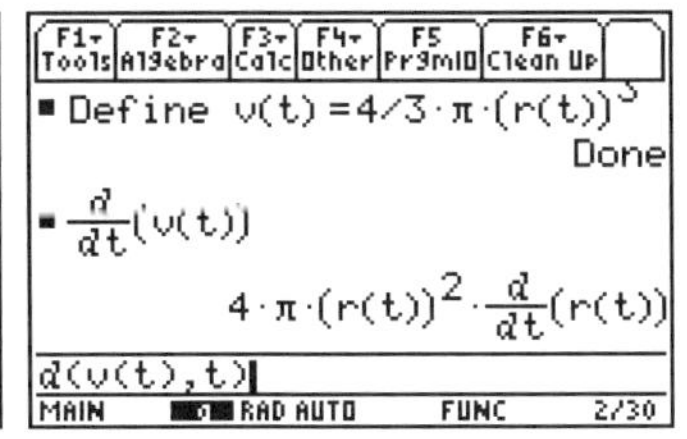

4·drdt·π·(r(t))^2
■ solve(4·drdt·π·(r(t))^2 = -
drdt = -3 / 2·π·(r(t))^2
...drdt*π*(r(t))^2=-6,drdt)
MAIN RAD AUTO FUNC 4/30

6.

2·π·(r(t))^2
■ drdt = -3 / 2·π·(r(t))^2 | r(t) = 1
drdt = -3 / 2·π
...-3/(2*π*(r(t))^2)|r(t)=1
MAIN RAD AUTO FUNC 5/30

Chapter 4

1.

■ ∫(x / 1 + 4·x^2) dx
ln(4·x^2 + 1) / 8
∫(x/(1+4x^2),x)
MAIN RAD AUTO FUNC 1/30

2.

■ ∫(a·b^(k·t)) dt   a·b^(k·t) / ln(b)·k
∫(a*b^(k*t),t)
MAIN RAD AUTO FUNC 1/30

3.

■ ∫ from 0 to π/3 (sin(x)·cos(x)) dx   3/8
∫(sin(x)cos(x),x,0,π/3)
MAIN RAD AUTO FUNC 1/30

4.

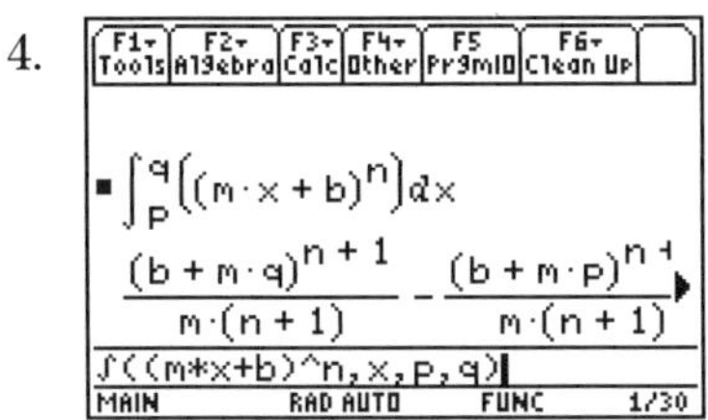

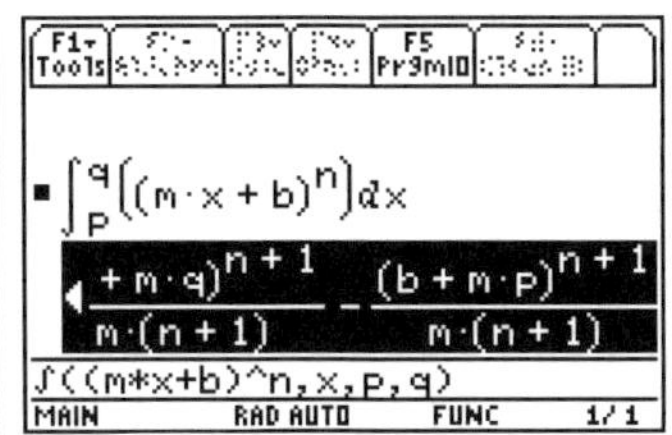

Chapter 5

1. **lram** = 4.9375, **rram** = 6.4375, **mram** = 5.65625
2. Same as Exercise 1.
3. **lram** = **rram** =**mram**= 17/3
4. Same as Exercise 3.
5. **lram** = $\frac{\pi}{4}$, **rram** = $-\frac{\pi}{4}$, **mram** = 0
6. **lram** and **rram** same as Exercise 5; **mram** = 2.356E-14 (in **APPROXIMATE** Mode)
7. The TI-89 cannot evaluate these limits.
8. 0

Chapter 6

1. .9471
2. 3.8202
3. $y = 2x\cos(x) + (x^2 - 2)\sin(x)$
4. $v = v_0 e^{-\frac{kt}{m}}$
5. $y = 19 - 16t^2$
6. $q = \left(\frac{-\cos(t)}{2} - \frac{\sin(t)}{2}\right)e^{-t} + \frac{1}{2}$
7. (a) $y = .8649 - 4.9t^2$

   (b) 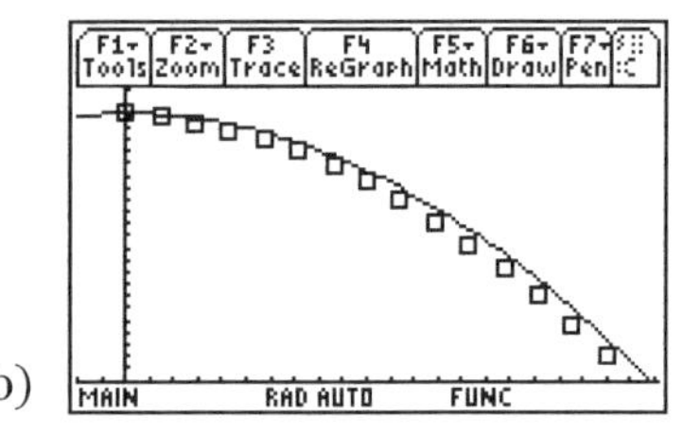

8. About 47°C
9. About 15.7 seconds
10. About 18 days

Chapter 7

1. 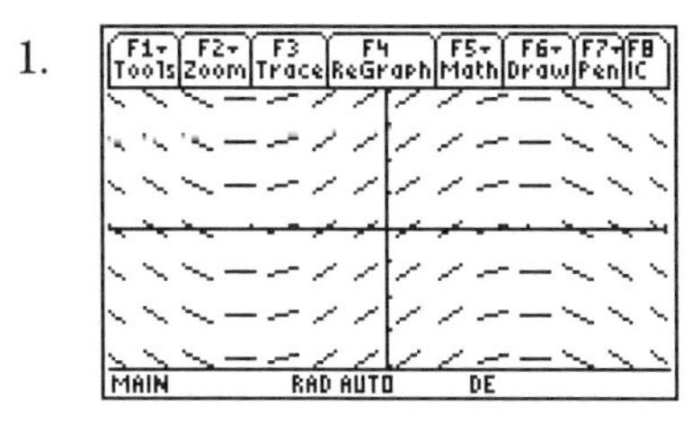

$[-\pi,\pi]$ x $[-1,1]$

2. 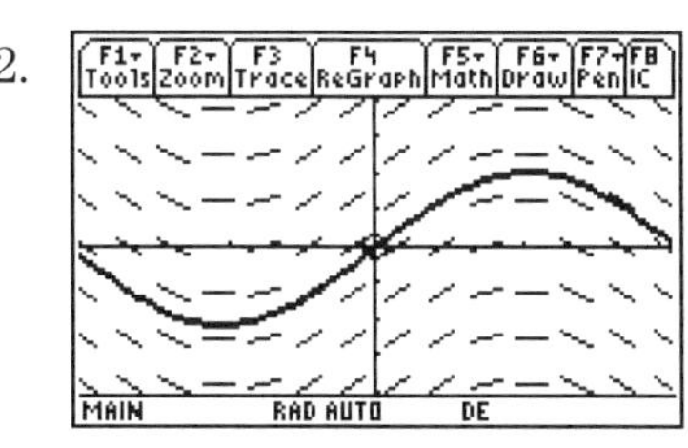

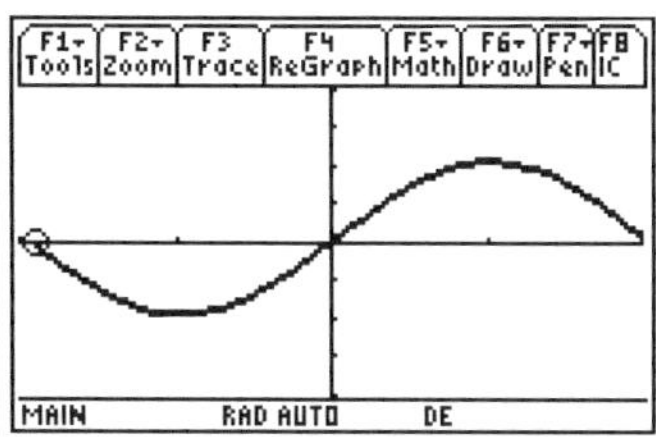

$[-\pi,\pi]$ x $[-1,1]$

3. 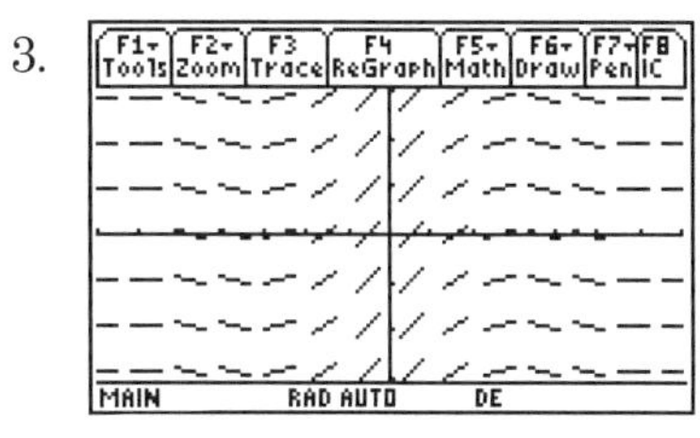

$[-7,7]$ x $[-3,3]$

4. 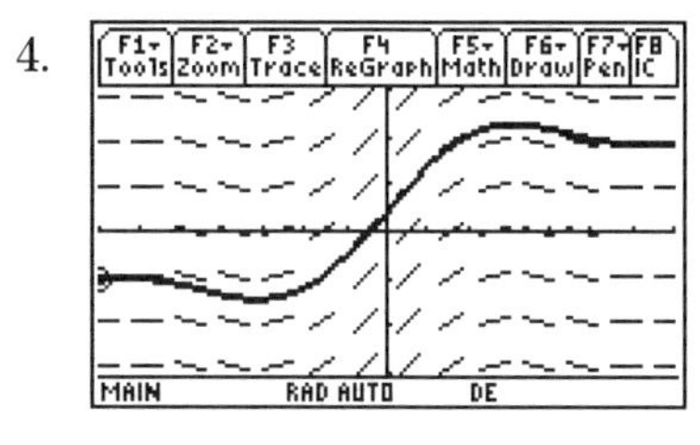

$[-7,7]$ x $[-3,3]$

5. 5700 years

6. about 13,200 years

7. 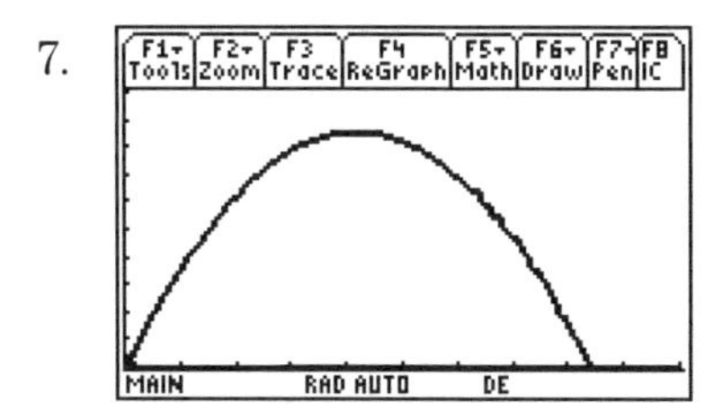

$[0,10]$ x $[0,100]$ $y = 41x - 4.9x^2$

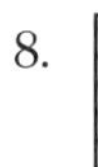

8. 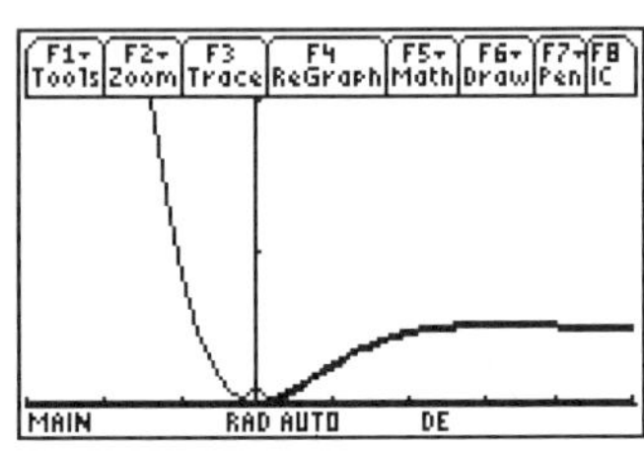

[-3,5] x [0,2] $y = \left(\frac{-\cos x}{2} - \frac{\sin x}{2}\right)e^{-x} + \frac{1}{2}$

9. About 379 feet

10. No

Chapter 8

1. (a) 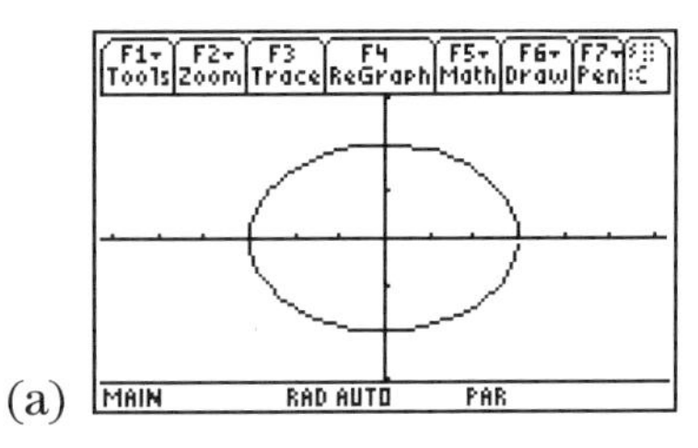

   (b) .667

   (c) $-\frac{2}{3}$

   (d) 15.8654

2. (a) 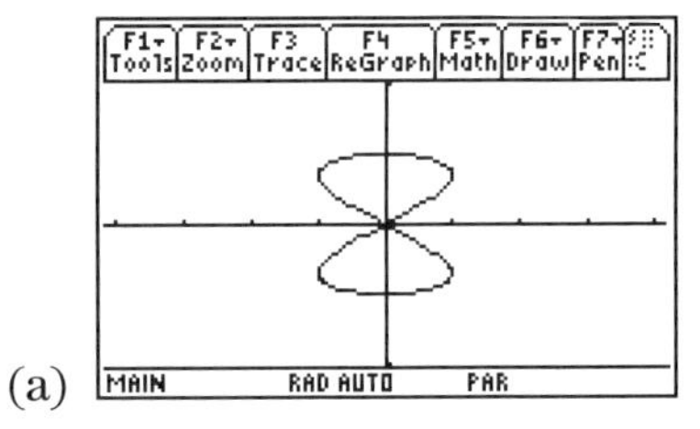

   (b) 0

   (c) 0

   (d) 9.42943

3. About 389 feet

4. 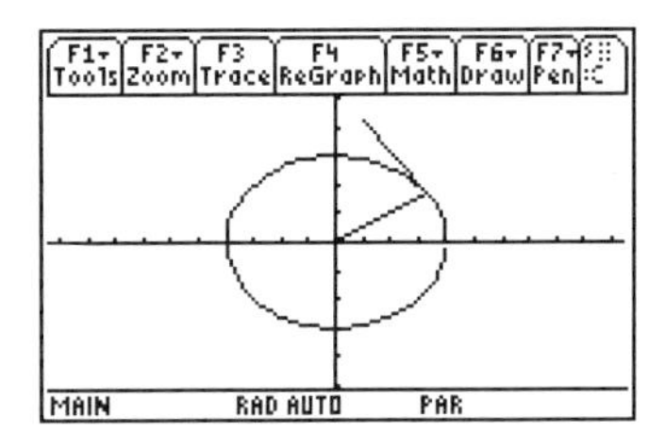

5. 1.921

6.

(a) 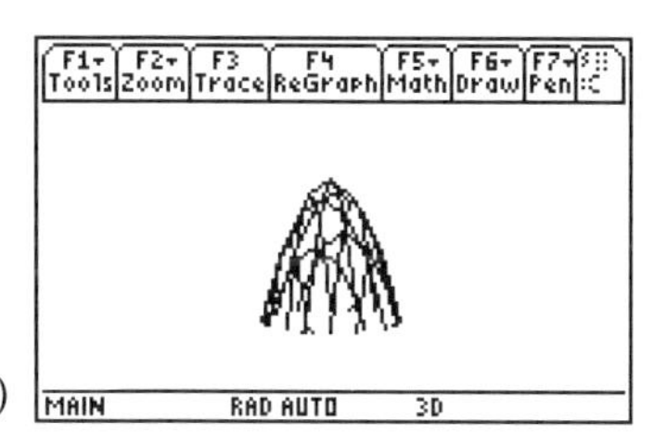

standard window

(b) 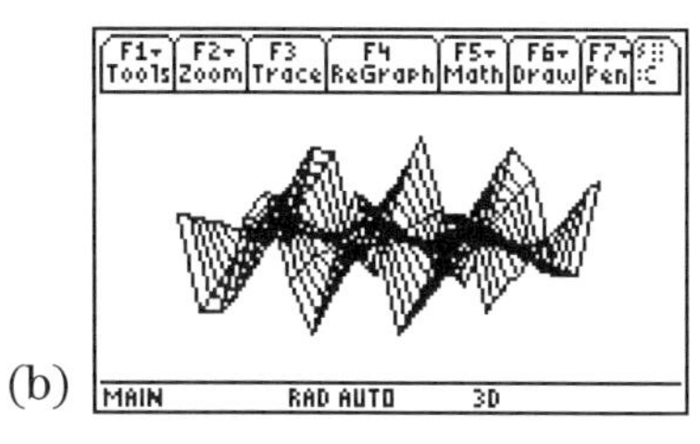

standard window

(c) 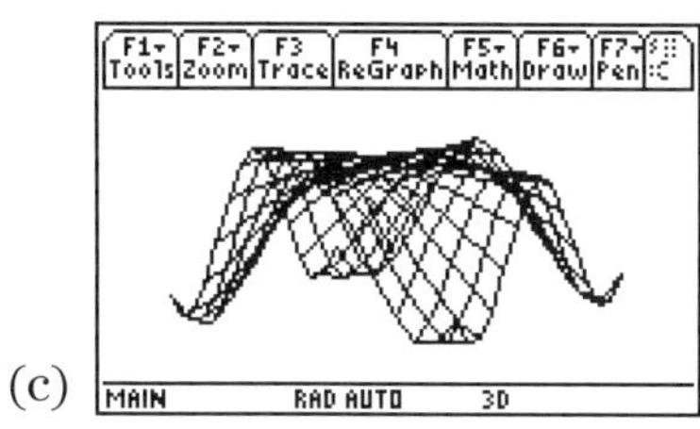

[-2,2] x [-2,2] x [-1,1]

Chapter 9

1. Converges to 0.
2. Converges to 0.
3. Diverges.
4. Converges to -0.693 (ln(1/2)).
5. Converges to 2.718 (e).
6. Converges to 3.141 (π).
7. Converges to 2.
8. Diverges.
9. $\frac{x^8}{315}-\frac{2x^6}{45}+\frac{x^4}{3}-x^2+1$
10. Same result as Exercise 9.
11. $\frac{(x-1)^5}{5}-\frac{(x-1)^4}{4}+\frac{(x-1)^3}{3}-\frac{(x-1)^2}{2}+(x-1)$
12. $\frac{(2x-\pi)^4}{384}-\frac{(2x-\pi)^2}{8}+1$

Index

—N—

—P—

—Q—

—R—

—S—

—T—

—V—

—W—

—Y—

—Z—